Concurrent Engineering
- the Agenda for Success

C³ INDUSTRIAL CONTROL, COMPUTERS AND COMMUNICATIONS SERIES

Series Editor: **Professor Derek R. Wilson**
University of Southern Queensland, Australia

Concurrent Engineering
- the Agenda for Success

Edited by

Sa'ad Medhat

*Dubai Polytechnic, Dubai Chamber of Commerce & Industry,
United Arab Emirates*

formerly at Bournemouth University, UK

RESEARCH STUDIES PRESS LTD.
Taunton, Somerset, England

JOHN WILEY & SONS INC.
New York · Chichester · Toronto · Brisbane · Singapore

RESEARCH STUDIES PRESS LTD.
24 Belvedere Road, Taunton, Somerset, England TA1 1HD

Marketing and Distribution:

Australia and New Zealand:
JACARANDA WILEY LTD.
Sydney Office, Suite 4A, 113 Wicks Road, North Ryde, NSW 2113, Australia

Canada:
JOHN WILEY & SONS CANADA LIMITED
22 Worcester Road, Rexdale, Ontario, Canada

Europe, Africa, Middle East and Japan:
JOHN WILEY & SONS LIMITED
Baffins Lane, Chichester, West Sussex, UK, PO19 1UD

North and South America:
JOHN WILEY & SONS INC.
605 Third Avenue, New York, NY 10158, USA

South East Asia:
JOHN WILEY & SONS (ASIA) PTE. LIMITED
2 Clementi Loop #02-01
Jin Xing Distripark, Singapore 129809

Library of Congress Cataloging-in-Publication Data
Concurrent engineering : the agenda for success / edited by Sa'ad
   Medhat.
         p.   cm. - - (Industrial control, computers, and communications
   series ; 13)
   Includes bibliographical references and index.
   ISBN 0-471-96144-2 (John Wiley). - - ISBN 0-86380-192-7 (Research
Studies Press)
   1. New products.   2. Concurrent engineering.   3. Production
engineering.   4. Production management.   I. Medhat, Sa'ad.
II. Series.
TS170.C66   1997
658.5 - - dc21   97-17286
                                                                          CIP

British Library Cataloguing in Publication Data
A catalogue record for this book is available from the British Library.

ISBN 0 86380 192 7 (Research Studies Press Ltd.) *[Identifies the book for orders except in America.]*
ISBN 0 471 96144 2 (John Wiley & Sons Inc.) *[Identifies the book for orders in USA.]*

Printed in Great Britain by SRP Ltd., Exeter

Series Editor's Foreword

Sa'ad Medhat - IBM Professor of Concurrent Engineering - has made a significant contribution to the introduction of Concurrent Engineering techniques in many industries, and through his leadership he has persuaded the Authors in this book to share their experience and expertise with a wider audience, through Research Studies Press. Sa'ad has drawn on the skills of application innovators across the world in order to present a state of the art 'statement of capability' in a topic that is now mandatory for commercial success.

In the energy-driven economy of yesteryear it was sufficient to manufacture goods almost on a trial and error basis. Demand was such that you could sell virtually everything, and even poor quality goods found a market. The advent of relatively cheap and abundantly available computer power has created a situation such that it is no longer 'can you make' but, rather, 'what do you want to make' that satisfies customer needs and aspirations. This transformation in consumer dynamics has been satisfied by the emergence of Concurrent Engineering as an essential applied technology. Initially, disparate techniques were developed such as Computer Aided Design (CAD), Computer Aided Manufacturing (CAM), Computer Aided Engineering (CAE), Knowledge Based Systems (KBS), etc. etc., but it is the integration of these techniques, through the company, and particularly coupled with market intelligence, that enables the concurrent development and production of new products to a specific specification, and to a delivery date, to be achieved. Without Concurrent Engineering, manufacturing companies cannot succeed in modern competitive economies.

Sa'ad Medhat has now taken up an appointment as Director of the Dubai Polytechnic and it is a particular pleasure to thank him and all his Authors for sharing their experience, wisdom and expertise by making it available to us through this book. On your behalf, the readers, I want to thank all the Authors for undertaking this significant task and for enabling us to participate and share the benefits of Concurrent Engineering.

Derek Wilson
Toowomba, April 1997

Preface

Concurrent Engineering has expanded and developed to reach out beyond simply looking at smarter ways for the design and production departments in a company to work more effectively together. Today, customer requirements are being met by many competitive suppliers. What is needed is a "cradle to grave" perspective for product development management and new technology introduction. Other disciplines are converging with Concurrent Engineering in response to the need for organisations to meet their ever increasing competitive challenges. Reducing time-to-market and product introduction costs are joined by the need to examine the dynamics of change within the organisation, and managing technology and processes. Technologies such as computer-based multimedia and client/server communications systems are now being adopted and introduced into the Concurrent Engineering communication "tool kit" for improved co-operative working and alternatives to co-location.

This book reflects these developments, with a major focus on organisational change to achieve more competitive product development. We have presented here a compilation of papers that brings together the rationale, planning and implementation of Concurrent Engineering aimed at the broad spectrum of industries that design, manufacture and deliver products.

The reader is taken through the industrial and commercial issues that have spawned Concurrent Engineering, the critical role of the customer and an examination of how Concurrent Engineering can be successfully implemented.

A portion of the book addresses how companies can involve their customers in defining product specification and ensuring that customer expectations are surpassed by the finished product. In an ever more vigorous competitive environment, maintaining customer interaction and loyalty is a critical paradigm for the future.

A company wishing to embark on the route of Concurrent Engineering needs to start with its overall corporate strategy which has been devised to achieve the company's stated aims of its mission or purpose. Whilst discussion on formulation of mission statements and their strategies lie outside the scope of this book, successful companies around the world have set their sights on a customer centred approach, arguing that development of market share is through long term customer loyalty, hence leading to a sustained superior performance for all its stakeholders.

Thus, for Concurrent Engineering to be a winning formula for successful product management throughout its life cycle, it must carry customer "advocacy". Team work is highlighted in several chapters in this book as one of the critical success factors for effective Concurrent Engineering. The challenge, however, is **how** to appropriately include the customer in the "team". One "customer presence" test to ask when making a product decision is - "Would the customer translate that decision, with its inherent trade-offs, as being to the customer's benefit?" If not, think again!

It is this customer centred approach that will drive innovation and sustainable growth, rather than developing a solution looking for a problem.

Finally, while this book draws mainly from the experiences of the larger multi-national company, the approaches to management of the issues and the process of resolution apply equally well to the small firm. Concurrent engineering is about how to respond to competitive pressures - faster and at a lower cost - with long term mutual benefit between themselves and the market they serve. All companies, I hope, share this vision.

Sa'ad Medhat, PhD, MPhil, PgD, CEng, FIEE, FCIM, MIMgt
IBM Professor of Concurrent Engineering
Intergraph Professor of Electronic Design Automation
Dubai, March 1997

Contents

Contributors

Scott R. Angster
School of Mechanical and Materials Engineering, Washington State
University, Pullman, Washington, 99165-2920, USA

Barry M. Brooks
PA Consulting Group, Cambridge Technology Centre, Melbourn,
Hertfordshire, UK, SG8 6DP

Chris Burns
BTR Industries Ltd, Challenge Court, Barnett Wood Lane, Leatherhead,
Surrey, UK, KT22 7LW

Mario W. Cardullo, P.E.
Belfield Group Inc., 1114 North Pitt Street, Alexandria, VA22314-1455,
USA *and* Northern Virginia Graduate Centre, Virginia Polytechnic
Institute and State University, USA

Nick Cramer
GEC-Marconi Avionics, Airport Works, Rochester, Kent, UK,
ME1 2XX, *formerly at Thomson Thorn Missile Electronics Ltd, 120
Blyth Road, Hayes, Middlesex, UK, UB3 1DL*

Nosa F. O. Evbuomwan
University of Newcastle upon Tyne, Department of Civil Engineering,
Newcastle upon Tyne, UK, NE1 7RU

Stephen G. Foster
Computervision Ltd., Argent Court, Sir William Lyons Road, Coventry, UK, CV4 7EZ, *formerly at Vickers Shipbuilding and Engineering Limited, Barrow in Furness, UK, LA14 1AB*

John R. A. Greaves
Coopers & Lybrand, Harman House, 1 George Street, Uxbridge, Middlesex, UK, UB8 1QQ

Mervyn J. Hall
Siemens Plessey Systems, Vicarage Lane, Ilford, Essex, UK, IG1 4AQ

Sylvie Jackson
formerly at Purchasing and Logistics Services, The Post Office, UK

Sankar Jayaram
School of Mechanics & Materials Engineering, Washington State University, Pullman, Washington, 99164-2920, USA

L. Jawahar-Nesan
School of Engineering and the Built Environment, University of Wolverhampton, Wulfruna Street, Wolverhampton, UK, WV1 1SB

Alan Jebb
The Total Quality & Innovation Management Centre, Anglia Polytechnic University, Danbury Park Campus, Main Road, Danbury, Chelmsford, Essex, UK, CM3 4AT

Andrew M. King
Engineering Design Group, Department of Manufacture and Engineering Systems, Brunel University, Uxbridge, UK, UB8 3PH

Kevin W. Lyons
National Institute of Standards and Technology, Gaithersburg, MD 20899-0001, USA

Sa'ad Medhat
The Dubai Polytechnic, PO Box 1457, Dubai, United Arab Emirates,
*formerly at Bournemouth University, Poole House, Talbot Campus, Fern
Barrow, Poole, Dorset, UK, BH12 5BB*

John Paul
BTR Industries Ltd, Challenge Court, Barnett Wood Lane, Leatherhead,
Surrey, UK, KT22 7LW

Sarah Philpott
EDS, 1-3 Bartley Wood Business Park, Bartley Park, Hook, Hampshire,
UK, RG27 9XA, *formerly at Thomson Thorn Missile Electronics Ltd,
120 Blyth Road, Hayes, Middlesex, UK, UB3 1DL*

Andrew D. F. Price
Department of Civil and Building Engineering, Loughborough
University, Loughborough, Leicestershire, UK, LE11 3TU

Franz J. Rammig
Cooperative Computing & Communication Laboratory C-LAB,
Furstenallee 11, D-33102 Paderborn, Germany

Jim Rook
Department of Electronics, Bournemouth University, Poole House,
Talbot Campus, Fern Barrow, Poole, Dorset, UK, BH12 5BB

Sang Sivaloganathan
Engineering Design Group, Department of Manufacture and Engineering
Systems, Brunel University, Uxbridge, UK, UB8 3PH

Bernd Steinmüller
Cooperative Computing & Communication Laboratory C-LAB,
Furstenallee 11, D-33102 Paderborn, Germany

PART I

CONCURRENT ENGINEERING
- THE CHALLENGE

CHAPTER 1

The Way Forward -
Managing Technology and Processes

M. W. Cardullo, P.E.

1 INTRODUCTION

Technology consists of knowledge, actions, and accouterments. Management of technology's principal objective is a rational approach, proceeding from knowledge, actions and accouterments to useable technological advancements, which meets both societal and competitive needs of the enterprises in which these items are embedded. However, the question which we must ask is:

WHY TECHNOLOGY?

The best place to initiate this Socratic quest is in the basis of the word *technology*.

Technology is a general term for the processes by which human beings fashion tools and machines to increase their control and understanding of the material environment [1]. The term is derived from the Greek words *tekhnë*, which refers to an art or craft, and logia, meaning an area of study; thus, technology means, literally, the study, or science, of crafting.

Technological innovations on a global scale seem to appear at a rate that increases geometrically, without respect to geographical limits or political systems. However, politics play an important environmental role in technology management. These innovations tend to transform traditional cultural systems, frequently with unexpected social consequences. Thus technology can be conceived as both a creative and a destructive process.

Technology has been a dialectical and cumulative process at the center of human experience [2]. Warnings on the duality of technology, i.e. beneficial and destructive qualities, were observed in the 1950s. It was observed that many products of technology had both useful and harmful or destructive aspects.

However, it has been very difficult in practice to predict secondary effects of new technologies.

Technology has always been a major means for creating new physical and human environments and one of the principal drivers has been societal needs.

Another of the drivers of technology has been competitive needs of enterprises, both public and private. During the last decade and a half, the United States went from the country with the largest trade surplus to one of the world's greatest deficits. Prior to these decades the United States led the world in competitive advantage which was secured by its leadership in technology. There is a definite relationship between technology, technological advancements, and national and enterprise competitive advantages.

Thus we can state that it appears that technology exists due to both a response to societal and competitive needs within nations and enterprises.

According to Einstein:

> *"Der Herrgott ist raffiniert aber boshaft ist Er nicht"*
> ..... *"The Lord is subtle, but he isn't simply mean."*

This statement is important for the performance of fundamental research [3]. The environment of fundamental science and technology does not *move* to research, i.e., the laws of the universe can counter research efforts. Technological change does not just materialize without being managed. Technological change is driven by enlightened self-interest of either private individuals, organizations, governmental bodies or a combination of these agents of change.

The definition of a breakthrough according to Webster's dictionary is:

> *"A sensational advance in scientific knowledge in*
> *which some baffling major problem is solved."*

According to Martino [4] a technological breakthrough is:

> *" An advance in the level of performance of some*
> *class of devices or techniques perhaps based on*
> *previously utilized principles, that significantly*
> *transcends the limits of prior devices or techniques."*

This definition would imply that adopting a successor technique that has a level of inherent capability higher than the prior technology would be considered a breakthrough. Many of the *breakthroughs* which helped shape this century had a substantial prior history. As it has been stated by Einstein, he *"...stood on the shoulders of giants."* The publication of Einstein's paper in 1905 established the equivalence of mass and energy and led to the cascaded events chain which resulted in nuclear weapons and atomic energy and now to a major new industry - *environmental remediation.*

Breakthroughs, either scientific or technological have many precursor events. Monitoring for breakthroughs is a systematic means for identifying using them to anticipate breakthroughs.

2 SOCIOTECHNICAL FACTOR

According to Porter et al [5] society and technology are abstractions for a wide variety of entities with various levels of concreteness and aggregation. Porter describes a *sociotechnical system* as an open system whose elements include:

- Technological devices and principles
- Scientific knowledge
- Institutions
- Individuals
- Money
- Natural resources
- Values

The sociotechnical system is viewed as a delivery system [6]. The boundaries of this system are generally arbitrary. The sociotechnical change processes of this system include:

- Development of scientific knowledge
- Development of technological principles
- Development of prototypes
- Production and diffusion of technical devices
- Changes in characteristics and objectives of institutions
- Changes in characteristics and beliefs
- Changes in wealth and resources
- Changes in values

Accordingly:

- Development of scientific knowledge is not cyclic
- Institutional change does not necessarily maintain equilibrium
- Changes in the state of resources are not necessarily the result of conflict
- Sociotechnical change is more of a spiral process in which system changes adapt under causal influences (stochastic

resonance) of its elements
- This is a non-linear system which is capable of chaotic behavior under certain initial conditions, i.e. the phase space is complex

This system can take several states or outcomes:
- Stabilized
- Incremental change
- Discontinuous change

It is the conflicts between different goals/values/world views which may lead to incremental or discontinuous change.

The system consists of four basic elements:
- Inputs
  - capital
  - natural resources
  - manpower
  - tools
  - knowledge from basic and applied research
  - human values
- Institutions and Organizations - Modify and control output
  - public
  - private
- Processes - Institutions interact through
  - information linkages
  - market interactions
  - political interactions
  - legal interactions
  - social interactions
- Outcomes - Effects on social and physical environments
  - direct (intended)
  - indirect (unintended)

The system proposed by Porter et al can be useful for conceptualizing the development of a single technology and the kinds of sociotechnical changes that could possibly result. It must be considered that in many instances there are many of these *single* sociotechnical systems within a larger context of a technological industry or industries. However, this approach can be useful in providing a manager of technology a means to:

- <u>Map</u> key players who can affect development
- <u>Depict</u> essential vector of technological development
- <u>Identify</u> resource gaps
- <u>Emphasize</u> leverage points
- <u>Highlight</u> important technological enterprises

3 TECHNOLOGICAL AND COMPETITIVE ADVANTAGE

Organizations and nations are faced with growing national and international competitiveness that places a strong emphasis on the following component measures:

- Productivity
- Rapid product introduction
- Quality
- Reliability

Technology is considered a measure of national power and an implement of public policy. In these terms *high technology* has been defined by Porter et al as [7]:

> *"high-value-added products that tend to embody the current state of the art and have a large research and development - R&D - content."*

The enterprise is the center of activity in asserting national competitiveness because in our capitalistic democratic society it is the firm in most instances that must deliver the technology. The recent changes in nations with planned societies have also shown that the enterprise is more effective in delivery of technology, products and services than the state.

However, the acceleration of technological change has placed stress on managers of technology due to:

- Products that have decreasing market or operational life cycles
- Increased concerns for safety, environmental effects and societal impacts
- The need to deal with strategic and technical considerations simultaneously
- Different investment decisions

Management of technology for competitive advantage is directly linked to R&D. R&D must be directed toward technologies that have the highest probability to impact on an enterprise's prime markets, i.e. therefore *true* identification of prime markets is crucial to any R&D program. The selection of R&D projects must understand customer needs and market dynamics. The resulting technologies must move through the total *developmental-technological* cycle in the form of beneficial innovations. Technological change is one of the principal drives of competitiveness [8]. This competitiveness permeates the total technological process.

Technological change plays a major role in structural change of existing industry, as well as in creation of new industries. It also serves to change the relationships in an industry and also among nations. Enterprises which have installed an effective system for encouraging and managing technology will be in a better position to develop and provide products and services [9].

There exists a *social attractor* in the fact that there is a vicious circle of lack of technology and underdevelopment, a fact which many nations are facing. This *social attractor* can eventually result in a long lasting bifurcation amongst nations. In fact a similar *social attractor* exists within nations, even the United States, which is facing a bifurcation between a technically trained and untrained society.

4 LESSONS LEARNED

During this century scientific and technological developments have progressed at an ever-increasing rate. Lessons are learned from the progress and pitfalls that have been experienced. These *lessons learned* demonstrate the impact of the management of technology. These lessons learned can be divided into those dealing with:

- Scientific Developments
- Technological Developments

4.1 Scientific Developments

Nuclear Physics: The majority of the major events in the development of nuclear physics during this century illustrate the rapidity of change in scientific developments. The first event, i.e. the 1905 paper by Einstein was a scientific breakthrough and could not have been predicted. This purely analytical paper *triggered* the growth of nuclear physics in this century. However, it was based upon developments in mathematics and physics in the nineteenth century which Einstein brought together in a new theory on the equivalence of mass and energy, a theory that was a violation of acceptable scientific theory. Einstein *took a step* into the unknown but based upon a sound mathematical basis.

The discovery that not all chemically identical elements were physically identical by Boltwood in 1906 and McCoy and Ross in 1907 showed that the radioactive elements ionium and radiothorium were chemically identical with the element thorium, although they had different masses - these were termed *isotopes.*

The scientific developments progressed in 1919 with the first artificially induced nuclear reaction. However, the source of this energy required a new instrument, i.e. a mass spectrometer, which allowed researchers to determine the masses of atoms. In 1919 Aston described his mass spectrograph which made it possible to distinguish between and to measure the different atomic weights of the same element. After this, the developments then came more rapidly, especially with the development of electrical machines to accelerate particles to speeds that would split atoms so as to cause the artificial transmutation of elements, i.e. the true *philosopher's stone.* It is important to note that these machines worked on the same principles as the mass spectrograph. The first of these scientific devices was the electrostatic generator of Cockcroft and Walton. In 1930 Lawrence invented the cyclotron which accelerated protons along a circular and far longer path than the linear accelerator of Cockcroft and Walton. These developments led to Chadwick's discovery of the neutron in 1932. Throughout the 1930s the worldwide scientific community conducted numerous experiments in which they bombarded various elements with neutrons. In 1938 two German physical chemists, Hahn and Strassman, discovered they could split the uranium atom into smaller fragments that gave off considerable energy. The element uranium had been identified and named over two hundred years before and isolated fifty years later. Two refugee Austrian scientists working in Copenhagen showed analytically that Hahn and Strassman's experiment with the splitting of the uranium nucleus when struck by a neutron, splits into two nuclei of approximately equal weights, one of which was an unstable barium nucleus; at the same time a considerable amount of energy is released.

The next steps quickly led Bohn, Fermi and others to the concept of a chain reaction - the rest is technological application leading to the first atomic weapons and nuclear energy production. In summary it appears that a chain of scientific events in various locations and not developed under a single aegis is the form of this type of scientific development. The chain can be described as:

- Analytical conceptualization based upon prior knowledge and techniques
- Followed by experimentation
- Need for new form of instrumentation
- Followed by additional experimentation and analytical conceptualization
- Finalized concept which can lead to technological application

Quantum Physics and Relativity: In 1901 Planck presented his quantum mechanical theory which was developed to reconcile the observed distribution of energy in the spectrum of heat radiated from a hot *black body* problem with accepted theory. Planck found a formula that fitted experimental results. According to his quantum theory radiant energy can be emitted or absorbed in discrete units. Planck did not really think much of his theory; to him it was a mere temporary expedient. In 1905, Einstein published his theory of relativity. Einstein's theories rejected Newton's basic postulate of the absoluteness of space and time and of the axiom of conservation of mass. Quantum theory and relativity found no practical application for many years. However, analytical and experimental work continued in institutes and universities. During the same period, the rapid growth of the radio industry and its supporting electronic technology was occurring.

To provide the necessary research and developments for the growth of these industries, technologists were recruited from the new physicists and engineers who were familiar with the work derived from Planck's and Einstein's theories. Using these engineers led to a number of technological applications:

- Radar
- Television
- Transistors
- Lasers

Thus the event chain in this instance can be summarized as:

- Experiments which do not confirm accepted theories
- Analysis to reconcile experiments and accepted theories
- Supporting analytical theories independent of experiments
- Incorporation of theories and data into an academic knowledge base
- Trained scientists and engineers knowledgeable about scientific developments which initially appear to have little

practical application
- Eventual employment of knowledge base to solve technological application problems

Computer Science: Today, enterprises increasingly rely on information technology both to perform their day-to-day operations and as a source of new products and services. However, the genius is based upon developments in computer science. According to Cardwell, the scientific motives were:
- the solution of specific scientific problems
- to build as advanced a device as possible to *push* against the frontiers of knowledge.

Babbage in the nineteenth century proposed a machine specifically for scientific purposes. He proposed that a scientific computer had to deal with:
- negative and positive numbers
- complex mathematical functions.

However, nothing happened until the 1930s when Konrad Zuse in Germany, with his own resources, started to build a mechanical computer. In 1937, Aitken published a specification for a computer machine together with a survey of previous work in the field. During the same period Turing published his *On Computable Numbers,* a classic in its field which also showed that there were insolvable problems. Turing conceived the abstract concept of a *universal computer.* The beginning of World War II accelerated the developments of computer science.
- In Germany, Zuse, now supported by the German government, built an electromechanical computer for ballistic calculations.
- In England, Turing and von Neumann working on the solution of the *Enigma code* problem built an early electronic machine based upon the binary concept of von Neumann.
- In the United States, the University of Pennsylvania in cooperation with the U.S. Army was working on an analogue computer for ballistic tables. A member of the staff, Mauchly, thought of replacing the mechanical analogue computer with a digital electronic computer. Mauchly and Eckert (also a member of the staff) built the first electronic computer, Electronic Numerical Integrator and Calculator ("ENIAC") in 1943.

Thus in this instance the demarcation between scientific and technological application is not as distinct. In this case we postulate the following chain scenario for the development:

- Conceptualization of scientific device but supporting theory and technology does not exist or is in an embryonic form.
- Gap in scientific development as supporting theory and technology proceed along independent paths.
- Major societal problem, i.e. war, requires a device to meet particular requirements.
- Scientific personnel using background and scientific knowledge incorporate concepts to formulate a dual use device.

Material Science: One of the most important breakthroughs in physics has had a profound impact on a particular branch of material science, i.e. superconductors. In 1911, the Dutch physicist Onnes discovered that electrical resistance in mercury vanishes when it is cooled to temperatures close to absolute zero (-460°F or 0°K). Onnes termed this phenomena super-conductivity. Other experimenters soon discovered that other metals and alloys become superconducting at very low temperatures.

The temperature at which a metal becomes superconducting is called the critical temperature. The higher the critical temperature of a metal or alloy, the easier it can be used in technical applications.

In 1957, 46 years after the discovery of superconductivity, Bardeen, Cooper and Schrieffer proposed a theory explaining the phenomenon. However, it was not until 1986/87 that two IBM researchers, Müller and Bednory developed materials that become superconducting at substantially higher temperatures than liquid helium. The new ceramic materials become superconducting between 90°K and 120°K, i.e. above the boiling point of liquid nitrogen, which is inexpensive and easy to maintain. However, in 1988 no satisfactory theory existed to explain the mechanism of superconductivity in ceramic materials.

The future widespread use of superconductors still awaits the development of a technologically usable material which is reliable and cost effective. This class of scientific development seems to follow the following chain scenario:

- Experimental work demonstrates a phenomenon which has no associated theory and which would be difficult to implement in the form of a usable technological application.
- Eventually a workable theory is developed to explain the

phenomenon.
- New materials or devices are discovered which increase the capability of the phenomenon, but still the usability, capability, reliability and economics are not yet demonstrated to have widespread use.
- Technologist either must wait for better capabilities or develop ways to circumvent particular technological roadblocks, like better manufacturing techniques.

4.2 Technological Developments

Unlike scientific developments, technological developments are a problem-oriented activity. Scientific developments are concerned with increasing the knowledge base about a particular phenomenon. Technological developments are concerned with the solution to a particular problem or creation of an application to perform a particular function. However, not all answers or developments result in useful or acceptable solutions which move on to Martino's next level of development, i.e. widespread use.

Technology leadership exerted through *function* produces products with advanced performance or features [10]. If commercial benefits of technology innovation are to be retained by the development enterprise, the organization must exert both technology and market leadership.

Semiconductor Technology: The development of the PowerPC by the consortium of IBM, Motorola and Apple countered Intel's dominance on the microprocessor market. Similarly SEMITECH was formed to counter the dominance of the Japanese microcomputer industry. However, while SEMITECH consortia resulted in a major change in the dominance of the international microcomputer industry, the PowerPC development has seemed to flounder; the question we must ask is:

> *Why does one form of technological development succeed in one case and basically fail in another case?*

Power PC Lesson: The PowerPC project grew of a desire on the part of IBM, Motorola, and Apple to create a family of microprocessors with the potential for gaining a significant market share and capable of supporting aggressive development in the future [11]. The consortium hoped that a new direction in microprocessor architecture would not only cut into Intel's market share but change the nature of computing. The three members of the alliance agreed that the new architecture would be based on IBM's existing POWER ("Performance Optimization With Enhanced RISC") microprocessor series, modifying it, but ensuring backward compatibility for application programs. While it was agreed that backward compatibility with the POWER architecture was desirable, the change consisted of *an evolution, not a revolution* in the design; however the question of *compatibility* was not simple.

Whatever the architecture, a new chip has to have circuits that are faster, and thus smaller. The speed of microprocessors has been increased through the introduction of dynamic complementary metal oxide semiconductor ("CMOS") circuits. As the size of the circuits decrease, changes in physical parameters create design challenges. Electrical noise is another problem that is exacerbated by the use of smaller and faster circuits. This problem is more prominent as succeeding chips in the PowerPC series aim for increasingly faster clock frequencies. In this case three enterprises, IBM, Motorola, and Apple, each with different cultures and management styles have worked on these and other development problems.

For each POWER instruction, there was a choice: either to hardwire it into the chip or emulate it in software. In the end, the controversy came down to deciding which choice would be the most efficient or least painful. A more serious problem was that while incompatibilities between the instruction sets of the POWER and PowerPC architecture would be eliminated when programs were recompiled, this was not possible with vendor-supplied legacy programs.

Furthermore, there were no simple and objective ways to make decisions, a common situation in both software and hardware design. All members of the consortia were aware that any change would incur costs to IBM in compatibility fixes, whereas Motorola and Apple, since they were moving to an entirely new architecture were less concerned with backward compatibility. According to Lerner, the debates were driven simply by professional pride in what had been done in the past. The biggest technical challenges in the PowerPC were not in small but controversial changes to instructions, but rather in the move to the 64-bit architecture. The second major change was the introduction of multiprocessing capabilities.

In September 1993, IBM announced the RISC System/6000 model 250 series of workstations featuring the PowerPC 601 processor. In March 1994,

Apple announced its line of PowerPC based Power Macintoshes. However, by August 1995, the industry magazine *Information Week* headlined an article entitled "PowerPC Shorts Out - New chip fails to spark large sales, so vendors target core markets."

The PowerPC had been designed to help Apple enter the enterprise market and spearhead IBM's desktop reassurance. Instead, sales for their computers based on the new chip were not meeting even conservative sales expectations; according to an industry executive the "*Sales have been pretty abysmal.*"[12]
The question we must ask is why?

SEMATECH Lesson: The Semiconductor Manufacturing Technology, Inc. ("SEMATECH") was established in 1987 to reverse the loss to Japan of semiconductor manufacturing leadership.   SEMATECH was a unique experiment in American industry-government cooperation. The enterprise was given the national mission of quickly restoring world leadership in semiconductor manufacturing to the United States.

SEMATECH was incorporated with 14 high technology companies representing 80% of the national capacity for semiconductor manufacturing. The actual goal of SEMATECH was to improve the state of U.S. semiconductor manufacturing technology, especially improving the current generation of equipment and the equipment that would come into widespread use within 5 to 8 years [13].   The emphasis of SEMATECH has been in manufacturing technology, not the products made with that equipment. SEMATECH is a horizontal consortium focused on strengthening upstream suppliers for its member enterprises. This is relatively easy to achieve because SEMATECH's focus is not on its members' products, but rather on its suppliers of manufacturing equipment.   SEMATECH has indeed achieved its goal of reversing manufacturing semiconductor leadership.

SEMATECH has demonstrated:

- Government-sponsored R&D programs can have economic impact on the competitiveness of a critical industrial sector.
- Constructive co-operation amongst its members without inhibiting competition amongst its members by working on problems upstream from its members.
- Linking semiconductor manufacturers and semiconductor-manufacturing equipment makers accomplishes the positive attributes of Japan's keiretsu without the negatives.

The question we must ask is why SEMATECH
succeeds and PowerPC seems to be failing?

Possible Answer: Gover has synthesized a competitiveness-commercialization model which may give us the answer to the question of why one consortia succeeds and another fails [14]. The dominant themes from the SEMATECH lessons include:

- Customer needs and market forces must influence the direction of development.
- Co-operative working teams spanning research to manufacturing, preferably working together in a shared facility.
- Communications must be facilitated by all practical means.
- Non-duplication of research.
- Emphasis on today's problem product and process improvement drives commercialization.

The PowerPC consortia did not focus on all of the elements necessary for maximum competitiveness in high technology industries. The PowerPC consortia did not strive for all four elements of the competitiveness model.

Information Storage Technology: Information storage technology has moved from the early 1940s low density storage to the very high density devices currently available in 1995. The early computers used magnetic tape which was developed in Germany during World War II to replace magnetic drums for back-up storage. As computers became more complex and faster magnetic tape was not sufficient for rapid storage and retrieval. This led to the introduction of the Winchester disk drive technology in 1956.

In 1982, IBM introduced the Personal Computer ("PC") which started another cycle of growth. With rapid progress in computer technologies in general, and especially in PCs, the need for mass data storage devices rapidly multiplied in recent years. Since 1956, when the first disk drive was developed by IBM, disk drive technology has gone through a series of rapid transformations in form-factor resulting in various smaller sizes of drives. Even in the initial developmental period, at the same time, capacities and performance doubled almost every 2-3 years. Recording densities have also gone up by more than 1,000 times and data access performance has gone up by at least 15 times over the last two decades [15]. What made this progress possible is a combination of developments in:

- materials for heads and media
- electronics for data communications
- head-positioning systems
- innovative designs for spindle motors
- new manufacturing and process engineering technologies

The disk drive industry is both technology and market driven. The upstream markets of:

- PCs
- software development
- microprocessors

are important to disk drive technology and markets.

In order to fully understand the technology issues of disk drives, the following technological developments are important:

- heads and slider development
- media and substrates
- actuator and head suspension
- spindle motor and its control
- read/write electronics and data channels
- computer interfaces

There are also a number of competing technologies which can seriously impact the disk drive technology industry, including:

- massive storage capacities of new tape drives
- massive storage capacities of optical drives
- flash memory cards with tremendous access speed

The characteristics of disk drive technology industry are:

- Short product life cycle
  - associated cost for tooling, etc.
- Low automation
  - high production cost
- Vendor concentration
  - increased cost of inputs
- Small distribution market
  - high opportunity costs
- Fragmented management
  - high communication costs

- Lack of process technology
  - yield/scrap cost

A review of this technology industry shows that it is [16]:

- Technology-driven
  - new products
  - new process
- Market-driven
  - market choice
  - product selection
- Quality-driven
  - OEM customers
  - vendors quality
- Competition-driven
  - relentless fight for market share
- Entrepreneur-driven
  - principal visionary

The computer industry and associated technologies have moved through Martino's development phases to become widespread and have moreover become a technology industry that has turned its basic output products into commodities. This end point for a technology industry results in a continuous reduction of profit margins and at the same time fuels innovations in technology. The industry is globalized with facilities, markets, vendors and customers located around the world [17]. This globalization results in problems of management and international technology trade.

5 ENVIRONMENT

Technology is embedded within a societal and competitive framework. The management of technology is similarly embedded and impacted by the environments of this embedment. To understand the process within these environments a *systems approach* conceptualization is important.

5.1 Stages of Technology

No technology goes directly from its creator to immediate application. It passes through a number of stages which represent greater degrees of practicality or usefulness. The following is a general set of stages in this process of creation to application [18, 19]:

- Scientific findings
- Laboratory feasibility
- Operating prototype
- Commercial introduction or operational use
- Widespread adoption
- Diffusion to other areas
- Social and economic impact

Scientific Findings: In this initial stage, the pre-technology exists in the form of the scientific understanding of some phenomenon, the properties of some material, the behavior of some force or substance, etc. At this stage it is not capable of being utilized to solve a problem or carry out a function. Scientific findings represent a *knowledge base* from which solutions to specific problems can be achieved.

Laboratory Feasibility: At this point in the process, the scientific finding has been identified so it can be applied to the solution of a specific problem, and a laboratory model has been developed. It is clear that no natural or physical laws are violated by the application and that it is capable of performing the desired function or solving the problem of concern, but only under laboratory conditions. The application at this stage may be described as a *breadboard model*. Use outside of the test environment would likely not be possible.

Operating Prototype: The application, at this stage, has been built which is intended to function satisfactorily. During the prototype stage it must be capable of operating within its intended environment by its intended users. Software developers usually assign this the term, *Beta version*, i.e. Windows 95 had over 200,000+ *beta* testers.

Commercial or Operational: The commercial technology at this stage faces its *true* economic test, i.e. economic feasibility. This is a crucial period for any technology or product, i.e. *Edsel Test*. The "first or Version 1.0 production model" is often considered the point when the technology or product has reached this stage. Non-commercial technology, such as those developed and implemented by governmental agencies, is represented at this stage by its "first use" in its intended operational environment.

Widespread Adoption: At this stage a technology has demonstrated that it is technically and economically superior to whatever else was used in the past, e.g. automobile versus a horse drawn cart, and it replaces the prior technology rapidly. This stage varies with each technology and some technologies never reach this stage and maintain only a *niche* market position.

Diffusion to Other Areas: At this stage the technology has not only dominated the application area in which it was first adopted but has been adopted in other areas as well. A case in point is the first adoption of jet propulsion to military aircraft; then to the adoption by the commercial airline industry. If the technology supplanted some earlier technology, e.g. transistor for vacuum tubes, at this stage the technology would have been adopted for purposes to which the earlier technology was never applied (used now in appliances, vehicles, cameras and other applications).

Social and Economic Impact: When a technology reaches this level, it has changed the *world view*. Television, INTERNET and other communication systems have caused major social changes. Not all technologies reach this stage directly. However, each technology has both direct and indirect impacts which taken together can cause *cascaded events* or world view impacts.

5.2   Systems Approach to Managing Technology

The systems approach represents a new approach to managing technology. The *new* system or holistic way looks at the world in terms of sets of integrated relationships. The systems view gives us a way of looking at complex management problems. It is a mode of organizing existing findings in reference to the concept of systems, and systemic properties and relationships.

Successful managers of complex systems must possess:

- Understanding of the technology of their *business*
- Understanding of the *basic concepts of management*
- Interpersonal style which facilitates their ability to get things done through others
- Ability to conceptualize and to operate using a systems approach

Most realistic management problems involve systems and the way they change. The *systems* concept has had a substantial impact on both the planning and the implementing functions of management. A *system* may be defined as: *a collection of components, interacting for a purpose.*

In the early 1960s researchers developed ideas on what eventually became known as *system analysis*. This methodology viewed that problems needed to be seen in relationship to the underlying systems.

System analysis has evolved from the early methodology:

1960s Systems Methodology for Problem Solving
- Define objectives
- Design solutions to meet objectives
- Evaluate cost/effectiveness of solutions
- Decide on the best solution
- Communicate the system solution
- Establish performance standards

1970s - 80s Systems Methodology for Problem Solving
- Formulate the problem
- Gather and evaluate information
- Develop potential solutions
- Evaluate workable solutions
- Decide on the best solution
- Communicate the system solution
- Implement the solution
- Establish performance standards

The value of the systems concept to management of technology can be seen in terms of the elements of the managers job:
- Overall effectiveness
- Conflicting organizational objectives

The systems approach to management of technology decision making necessitates the use of objective analysis of decision problems. A systems approach to technology management problem solving requires having the perspective to deduce important system variables and the relationship between them.

A general statement of this approach is given by the following phases [20]:
- Creativity
- Choice
- Implementation

Creativity: There are various ways to assist technology managers creatively about enterprises. An enterprise can be viewed as:

- Closed system - as a *machine*
- Open system - as an *organism*
- Learning system - as a *brain* or *neural net*
- Emphasis in norms and values - as a *culture*
- Unitary political system - as a *team*
- Pluralist political system - as a *coalition*
- Coercive political system - as a *prison*

The view of the organization has an impact on its approach to managing technology. A combination of approaches or views, i.e. an organization which is open, learning and unitary in terms of application will likely result in maximizing the creativity phase.

Choice: The task during the *Choice* phase is to choose an appropriate system-based technology or technologies to suit particular characteristics of the problem revealed by the examination conducted in the creativity phase. The most probable outcome of the *Choice* phase is that there will be a *dominant* technology chosen to be modified or developed further in use as more information from the environment is obtained.

Implementation: The *Implementation* phase is responsible for taking the chosen technology and *bringing it to market* or utilization. During this general phase a number of actions are required by the organization.

The following chart maps the process of creation to application to the general system phases.

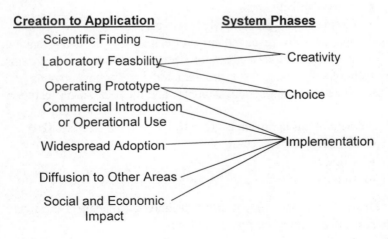

Creation to Application

Scientific Finding
Laboratory Feasbility
Operating Prototype
Commercial Introduction or Operational Use
Widespread Adoption
Diffusion to Other Areas
Social and Economic Impact

System Phases

Creativity
Choice
Implementation

5.3   Precursors of Technological Change

The precursors of technological change are [21]:

- *Incomplete* inventions - These may demonstrate the feasibility of some technology but may require other elements before they can be deployed economically or in a particular environment.

  Example: Controlled fusion

- Supporting technologies
  - Performing improving technologies needed by the basic technology.

    Example: Lasers for high capacity fiber optics for wide band communications
  - Cost-reducing technologies needed by the basic technology.

    Example: High volume/high quality integrated circuit manufacturing techniques which were need to make low cost personal computers possible

- Complementary technologies - A technology which the basic technology must interact with in order to be useful.

  Example: Satellite position systems in order to develop electronic highway control

- Prestige or high performance application - Use of a technological advance in a prestige or high performance application before transition into general use.

  Example: Use of technology in race cars before introduction into automobile production

- Incentivization - A requirement of an incentive for using the technology such as reduced cost or elimination of externalities.

  Example: Clean air act leading to clean coal technologies

5.4   Private Sector

In a societal sense, a business enterprise is expendable. Customer satisfaction usually determines enterprise survivability. This is in contrast to a governmental or non-profit organization such as a hospital or university which usually remains in business no matter how poorly they satisfy their customers.

If a company wants to stay in business, it must continue to satisfy its customers. It must anticipate changes in their wants and needs, as well as changes in the ways these wants and needs can be satisfied. A new company not having customers must satisfy its *stakeholders*, i.e. banks, investors, principal staff who have invested their time, creativity and energies in an unproven enterprise.

An important consideration in the success of a business is to accurately identify what business the enterprise is in. This is usually done in only superficial terms which often are too narrow, leaving the enterprise vulnerable to technological shifts and changes in customer needs.

The nature of an enterprise's business can be obtained by observing:

- Specific function performed by the enterprise
  Example: Banking
- Specific product or service provided
  Example: Steel manufacturing
- Specific process utilized
  Example: Communication industry
- Specific distribution system utilized
  Example: Mail order companies
- Specific set of skills
  Example: Consulting company
- Specific resource utilization
  Example: Paper manufacturer-uses wood pulp and
  recycled paper

Once the type of business is determined it is possible to evaluate the consequences of technological change. It is possible that technological change can actually lead to the disillusionment of the enterprise. Therefore, it is important that managers be very alert to technological changes that can alter the way they do business. These changes, if not anticipated, may leave the enterprise unable to compete effectively with companies in the same business that did anticipate the changes properly. An example is Wang Computer and Digital Equipment Corporation - one company went into bankruptcy and the other lost significant market share because they both failed to adequately see the changes by the small personal computer on corporate buyers based upon the Intel chip technology.

Technological change can alter the fundamental nature of an enterprise. These impacts can occur in:

- Function - An enterprise is vulnerable either to a technological change that makes the function unnecessary or to one that performs the function in some other manner.

  Examples:
  - Transportation: Roll-on-roll-off trucking
  - Utilities: electric versus gas lights
  - Shipping: Container shipping

- Product - A technological change may allow some different product to be used for the same purpose.

  Examples:
  - Transistor versus vacuum tubes
  - Automobile versus horse drawn carriages
  - CD ROM versus printed material

- Process - A new process for manufacturing a product providing some service that can almost directly replace an older process.

  Examples:
  - Federal Express versus U.S. Postal Service
  - Continuous casting of utility wire versus welding sections
  - Bioprocessing versus chemical processing
  - Direct coal injection for steel manufacturing versus the use of coke

- Distribution - Technological change has altered the way many firms distribute their products.

  Examples:
  - Airfreight versus trucking and ocean shipping
  - INTERNET versus printed media

- Skills - The skill mix in an enterprise may drastically alter with technological change. This can occur because of new processes or new products.

  Examples:
  - Computer controlled manufacturing versus manual control
  - Computer type setting versus individual type setter

- Raw Materials - New technology can result in a change in material used for both process and products. Man-made materials in many instances have replaced natural materials.

  Examples:
  - Reinforced plastic versus metals in cars
  - Composites versus metals
- Management - The manner in which a firm is managed can also be impacted due to technological changes.

  Examples:
  - Groupware versus individual products
  - Virtual organization versus physical organization
  - Management sciences versus solely managerial judgement
- Support - Operational activities such as hiring, payroll and other support functions can be impacted by technological change.

  Examples:
  - Behavioral technology versus standard hiring practices
  - Management Information Systems versus manual payroll and record keeping.
  - FAX versus phone support

However, technological change can have an impact on one or more of these enterprise aspects including its managerial and support activities.

5.5 Governmental Sector

The structure of society requires governmental organizations to provide functions for the *good and welfare* of society. Thus, a stated goal of governmental technology policy is to make the most efficient use of taxpayer-supported science and technology in achieving national security and the national goals of increased U.S. competitiveness in world markets, improved quality of life for all Americans and continued economic growth.

This goal and the political environment in which it is pursued will often cause the management of technology to take on different aspects from that which occurs in the private sector for commercial purposes. For example, in the interest of national security and/or long-range economic competitiveness, the government may undertake costly, high risk research and development which is either not affordable by the private sector or represents excessive financial risk.

The importance of cost and risk considerations suggest the following overall structure for considering the management of technology in the federal government:

- Government Funded Mission and Regulatory Organization
- Revenue Generating Government Owned Enterprises

Government Funded Mission and Regulatory Organizations: The basic difference from the private sector is that the objectives of this government sector are established outside of the organization and are characterized by the absence of a profit motive. Similarly their budgets for accomplishing these objectives are determined by the legislature as are the staffing levels.

Within this government sector there is a great deal of variation. The defense/space organizations are high technology organizations which manage and implement the full technology cycle. The cycle encompasses the following three phases [22]:

- Basic Research - Discovery of fundamental new principles about natural phenomena. The objective of basic research is to increase the knowledge base of physical phenomena and processes. This phase involves:
  - Synthesis of hypothesis
  - Theory
  - Observation

  This activity is usually done without any concern for practical applications. However, without basic research, technological progress would soon be reduced to chance. In some instances invention has proceeded scientific explanation. Basic research seems to thrive in an advanced academic environment.
- Applied Research - Known principles are exploited and useful applications can be foreseen. This is the invention and innovation stage and may include technology demonstration. However, applied research has definable and limited goals.
- Development - At this stage engineers usually replace the scientist. This is the stage where engineering and economics are used to determine whether a particular technological innovation can be *achieved* within economic constraints and where those proposals found to be technically and economically feasible are implemented. This is also known as the *hardware* stage.

- Operations - At the outset of this stage the system resulting from the development phase is accepted by the ultimate user who proceeds to deploy, operate and maintain the system for its intended purpose throughout its lifetime.

In executing this cycle, within the defense/space component of this sector, the government makes extensive use of industrial contracts to bring technology from conception to implementation, with industry playing the dominant role in development. By contrast, the non-defense/space mission and regulatory governmental organization research and development ("R&D") efforts are more concentrated on the pre-development phases of the technology cycle and a significantly higher percentage of the overall effort may be in governmental laboratories using governmental staffing.

One of the important issues is the relationship between defense/space spending and economic performance. Defense/space governmental R&D programs have supported many basic and applied-research studies in addition to many applied and development activities. There appears to be direct and indirect linkages between defense/space R&D, technology and economic performance [23]. These include:

- Direct effects on private sector technology, technical-skills formation, and overall economic performance.
- Indirect effects on private sector economy via its effects on technical change.
- Indirect effects on the private sector on technical-skills formation.

With the decline of the Soviet threat, the urgency of domestic problems, and the budget deficit, public pressure has intensified for the defense/space research and development community to produce technologies with commercial applications [24]. Analysis of time-series data on U.S. defense R&D activities by Charabarti and Anyonwu provides some evidence of the importance of governmental R&D on civilian economic performance in an indirect way. The effect is observed particularly through change as measured by the number of patents granted to organizations and individuals. Chakrabarti and Anyonwu argue that they believe that patents are the best available indicator for technology. Data also indicates that government promotes R&D investment by awarding major contracts through a competitive procurement process. By this process the government reveals the demand for certain technological innovations and encourages the private sector to invest in R&D. Chakrabarti and Anyonwu also observed that non-R&D aspects of defense spending appeared to have *no statistically significant* effect on both of the major components of civilian economic performance, i.e. technological change and technical skills formation.

Increasing the integration between defense/space and civilian technology could possibly lower governmental cost, promote increased private sector use, increase available industrial capacity, and likely strengthen national security. These possible outcomes are yet to be verified. Nevertheless, the defense sector has begun placing increased emphasis on research and development aimed at technologies that may enhance industrial competitiveness over both the long and short term.

This change of a policy paradigm has been accelerated by a number of factors including the following statement by the Clinton Administration [25]:

> *"American technology must move in a new direction to build economic strength and spur economic growth. The traditional federal role in technology development has been limited to support of basic science and mission-oriented research is the Defense Department, NASA, and other agencies. We cannot rely in the serendipitous application of defense technology to the private sector. We must aim at these new challenges and focus our efforts on new opportunities before us, recognize that governments can play a key role helping private firms develop and profit from Innovation."*

This raises certain questions as to the process of trying to refocus defense/space technological activities and how these activities are managed.

Revenue Generating Government Owned Enterprises: While units such as the U.S. Postal Service, Uranium Enrichment Corporation, municipal utilities and other Government Owned Government Operated ("GOGO") organizations do not have a profit incentive similar to the private sector, they do have measures of economic effectiveness and competitors. In this sense the management of technology within these organizations is similar to the private sector.

5.6 Management of International Technology

The economy of the world is being transformed by a technological revolution.

The transformation emanates from the information technology section [26]. The transformation is international in nature and therefore much of the adjustment in national economies is transmitted through the international trading system. No comprehensive data are available that identify all international technology flows. A great deal of this trade is in the form of R&D intensive exports, in which technology is embodied.

It is therefore not surprising to recognize that technological developments have an impact on international policy in both economic and political-military spheres.

Technology moves across national boundaries in a number of ways:

- International Technology Market - Independent buyers and suppliers.
- Intra-firm Transfer - Transfer takes place through either a joint venture or wholly owned subsidiary.
- Government Agreements and Exchange - Counterparts can be either public or private actors.
- Education, Training and Conferences - Dissemination of information is made public for common consumption by either a general or specialized audience.
- Pirating or Reverse-engineering - Access to technology is obtained while resort to market is avoided but at the expense of the proprietary rights of the owner(s) of the technology.

International management of technology will be involved when the production plant is located in one country and some of the technological and managerial inputs to the investment process are imported from suppliers in another country. Similarly, the technological process may all be located in one country but the market is world-wide. Many multinational companies have plants in various countries with management and design functions in other countries. Communication through the use of advanced communication systems such as wide area networks ("WAN") provide the means for the enterprise to effectively operate.

According to Ohmae [27], most managers in companies that have operated internationally for years are *nearsighted*. Accordingly these managers may manage complex organizations with elements in a number of different countries, or have joint ventures, sources and sell all over the world, but their vision is usually limited to their home-country customers and organizational units that serve them.

International technological management requires additional talents and background from that required in a technology company only dealing in one country - a rapidly decreasing number. These characteristics include:

- Knowledge of international trade and the regulations covering these trades.
- Knowledge and sensitivity of the socio-economic environments in which they will operate.
- Knowledge of international financial instruments, currency and other economic considerations.
- Knowledge of the legal systems in which the enterprise will operate.

The elements of the management of international technology, where one country supplies the technology and the other produces, are [28]:

- Management and executive of R&D - This is performed by the suppliers of the technology.
- Management and execution of pre-investment and feasibility study - This task is performed either by the technology importer or consulting firm(s) hired by them.
- Management and execution of design and engineering services - These services are usually performed by the technology suppliers.
- Management and execution of capital goods production - Process management and total involvement of the employees are directly applicable to this phase. If the suppliers and the purchaser work as partners, better outcomes are expected as the purchaser has better knowledge of the local conditions including the needs and expectations of the ultimate consumer.
- Management and execution of installation and commissioning service - Leadership, effective communications, education and training, and process control are the important elements in this stage of the process.

The source of much of the material in this chapter is:
M.W.Cardullo, **Introduction to Managing Technology**, Research Studies Press Ltd, 1996

32

REFERENCES

[1]   "Technology," Microsoft ^(R) Encarta '95.,Microsoft Corporation, 1995.

[2]   Ibid.

[3]   Jantsch, Erick, *Technological Forecasting in Perspective*, OECD, Paris, 1961.

[4]   Martino, Joseph, *Technological Forecasting for Decision Making*, McGraw Hill, New York, NY, 1993.

[5]   Ibid.

[6]   Ibid.

[7]   Ibid.

[8]   Porter, Michael E., *Competitive Advantage*, Free Press, New York, NY, 1985.

[9]   Edosomwan, Johnson A., *Integrating Innovation and Technology Management*, John Wiley & Sons, New York, NY, 1989.

[10]   Ibid.

[11]   See Edosomwan.

[12]   Gillooly, Brian, "PowerPC Shorts Out," *Information Week*, pp. 24, August 21, 1995.

[13]   Gover, James E., "Analysis of U.S. Semiconductor Collaboration," *IEEE Transaction on Engineering Management*, vol. 40, pp. 104 - 113, May 1993.

[14]   Ibid.

[15]   Jabbar, M. A., "Globalisation of Disk-Drive Industry," *Proceedings of IEEE Annual International Management Conference*, June 28-30, 1995.

[16]   Ibid.

[17]   Ibid.

[18]   Bright, James R., "The Manager and Technological Forecasting," in James R. Bright (ed.), *Technology Forecasting for Industry and Government*, Prentice-Hall, Englewood Cliffs, NJ, 1968.

[19]   See Martino.

[20]   Flood, Robert L and Jackson, Michael C., *Creative Problem Solving - Total System Intervention*, John Wiley and Sons, Chichester, UK, 1991.

[21]   See Martino.

[22]   Prehoda, Robert W., *Designing The Future - The Role of Technological Forecasting*, Chilton Book Co., Philadelphia, Pa., 1967.

[23]   Chakrabarti, A. K. And Anyonwu, C. L., "Defense R&D, Technology and Economic Performance: A Longitudinal Analysis of the U.S. Experience," *IEEE Engineering Management*, vol. 40, pp. 136-145, May 1993.

[24]   U.S. Army, *Army Science and Technology Master Plan*, vol. 1 and 2, November 1993.

[25]   President William J. Clinton and Vice President Albert Gore, Jr., *Technology for America's Economic Growth, A New Direction to Build Economic Strength*, Feb. 22, 1993.

[26]   Preeg, E.H., "Who's Benefiting Whom? - A trade Agenda for High-Technology Industries," *IEEE-Engineering Management Review*, vol. 22, pp. 75-83, Summer 1994.

[27]   Ohmae, Kenichi, *The Borderless World*, Harper Business, 1990.

[28]   Quazi, Hesan A., "Application of TQM Principles in International Technology Transfer Process: an Integrating Framework," *Proceedings of IEEE Annual International Engineering Management Conference*, pp. 128-133, June 28-30, 1995.

CHAPTER 2

CE - Enabling Technologies

F. J. Rammig *and* B. Steinmüller

1 INTRODUCTION

Concurrent Engineering (CE) is a systematic approach to the integrated, concurrent design of products and their related processes, including manufacture and support. This approach is intended to cause developers from the outset, to consider all elements of the product life cycle from concept through disposal, including quality, cost, schedule, and user requirements. It follows that different goals and intentions within the entire design process are attacked simultaneously in an integrated manner, e.g. performance, manufacturability, testability, cost, reliability, serviceability, reusability, recyclability. As a consequence, CE is a multidisciplinary approach which in turn causes us to consider teamwork as a basic organizational principle (see also [1]).

The inherent complexity of CE leads to the conclusion that CE can only be implemented successfully on the basis of powerful enabling technologies as provided by Computer Aided Concurrent Engineering or "**CACE**" product data and organizing teamwork of heterogeneous, distributed teams. Moreover, specific engineering tools are required to allow the concurrent consideration of particular design goals through all phases of the design process, while multidisciplinary tools are needed for the integrated design of complex products.

This chapter gives an overview of computer aided enabling technologies for Concurrent Engineering, i.e. CACE-technologies with a focus on general services and infrastructures.

Section 2 introduces CE-teams and their main requirements, followed by an overview of basic architectural concepts and frameworks. Section 3 describes CE-middleware services, which as an intermediate support layer plays a prominent role in CACE-environments. Finally, in sections 4 and 5, domain-neutral and domain-specific CACE-applications are treated and illustrated by typical examples.

2 CE-TEAMS, REQUIREMENTS, BASIC ARCHITECTURES

2.1 CE-Teams and Users

Concurrent Engineering requires the interaction of multidisciplinary users organized in CE-Teams. Depending on the complexity and size of the engineering task these teams can become very large and can be geographically distributed. In fact, there is a recent trend to dramatically speed up the engineering process by distributing sub-teams around the world in such a way, that 24 hour continuous work becomes possible by shifting work from one team to another one at the end, or beginning, respectively, of a working day.

Within a CE-team, typically, members from the following disciplines can be found:
- product research, development, fabrication and maintenance
- product planning and marketing
- business administration
- methodology and process support
- hardware and software support and system administration

Thereby the first three groups primarily are the **end-users** of CE-enabling technology, while the latter two are **support-users**, who are responsible for installing, maintaining or building the enabling technology in such a way as to make the end-users most productive [2, 3, 4].

2.2 General Needs and User Requirements

Taking into account the heterogeneity of CE-users and teams, it is clear that a broad spectrum of diverse requirements has to be satisfied. Looking at the end-user needs at first, the following main requirements with respect to CACE-technologies can be identified:

1. CE-Teams
   - management of distributed, multidisciplinary CE-teams
   - management of roles, relations and access rights
   - provision of efficient communication and co-operation tools
2. CE-Processes
   - management of distributed engineering activities and workflows
   - provision and management of goal driven design tools (design-to-X)
   - provision and management of multidisciplinary design tools
   - tool-interoperability and inter-tool communication
3. CE-Data
   - management of heterogeneous, distributed database systems
   - shared storage of fine and coarse granular data and interrelations
   - administration of versions and configurations
   - provision of standard exchange interfaces, access and retrieval tools

As far as the "support-users" are concerned, the following additional requirements can be identified:

4. CACE-Technology Installation and Configuration
   - automatic installation procedures
   - platform independence, portability
   - customizability
5. CACE-Technology Creation and Extension
   - openness, standardized interfaces
   - generators and scripting languages
   - modular architectures
6. CACE-Technology Maintenance
   - diagnostic tools
   - distant automated support

Of course, this list is far from complete. There are many additional requirements specific to the respective engineering task and application domain at hand [3-7]. Some of these additional requirements will be touched upon in the sequel of this article in the appropriate contexts. The top-level requirement pertaining to conceptual integrity, integration and cohesion will be introduced in the following section.

36

2.3 Basic Architectural Concepts and Requirements

Considering the diversity of tools and technologies offered to the solution of the CE-problem, CE-users and teams would like to work within an environment which is integrated and complete, and offers all needed functionality in a uniform, consistent way throughout all engineering phases. In practice, unfortunately, this is not possible and can only be realized on the relatively low level of individual CAx-Systems which support selected engineering aspects with an integrated suite of tools. As soon as different parts of an organization and more complex tasks are involved, different systems come into play, which at best can be coupled or federated to deliver uniformity and consistent access for a limited number of functions. While a single organization still can exert some control over the systems to be employed, hardly any control can be exerted on the global scale and as a consequence, different environments are usually only loosely coupled. The latter level, however, becomes increasingly important for CE-teams co-operating and competing on an international scale.

As a consequence, when looking at architectures from a general organizational point of view, we find a situation as sketched in Figure 1.

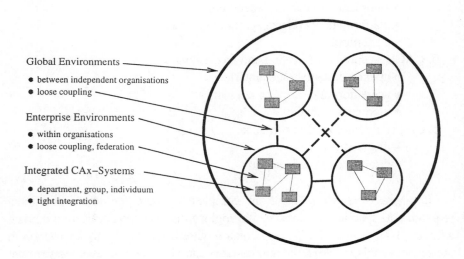

Figure 1   *A Global View on CACE-Technology - Three Main Levels of Clustering and Cohesion*

An important goal for future design and standardization efforts, is the question how more conceptual integrity and integration can be realized on the upper two environment levels without unduly impeding flexibility and openness for new solutions.

Considering the CACE-services needed for satisfying the diverse requirements mentioned in section 2.2, it is clear, that a rich set of domain-specific and domain-neutral end-user services or "applications" must be provided. In order to bridge the gap between the low-level services found on heterogeneous computer platforms and the high-level applications in an effective way, it is moreover necessary to provide services on an intermediary level, which complement and extend the basic facilities of typical platform services: "CE-Middleware". Thus, we arrive at the following conceptual CACE-architecture:

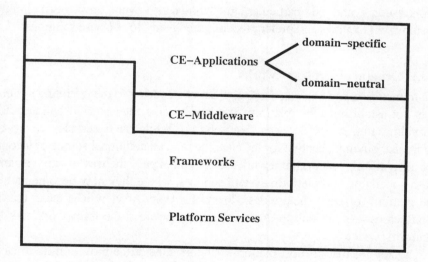

Figure 2   *A Functional View on CACE-Technology - Main Levels of Abstraction*

This architecture is a suitable means for ordering and classifying the large set of diverse CACE-services and reflects the principle layered structures found in real CACE-environments.

Obviously, middleware and domain-neutral services play an important role in other computer-based environments as well. Thus it is not astonishing, that a lot of effort has been invested in identifying those services which are the core to many different applications, and offering them as special entities called "frameworks". Thus, before turning to more specific CACE-services, we shall have a short look on frameworks.

2.4 Frameworks

Frameworks can be defined as a set of related computer-based services for building and maintaining computer-based environments. These services comprise middleware and domain-neutral applications and are typically based on a common architectural principle. Services for development and run-time support are usually provided in separate packages and toolkits. Thereby the development kit is of main interest to "support-users" who want to create specific customized environments by using the full power of the underlying framework, while the run-time systems usually are of interest to less ambitious support-users and in particular to "end-users" who only need limited customizing facilities and prefer preconfigured ready-to-use base solutions.

2.4.1 General object frameworks

As frameworks can be considered as a direct extension to the operating system, it is not astonishing that platform vendors became interested in building and providing framework services themselves. While such services create an additional buying incentive for the customer and an additional source of income for the platform vendor, framework services often pave the way to later versions of the operating and platform system services, where they may become part of the platform offering themselves. From this point of view it is clear, that in many cases it is difficult to draw a clear line between the framework and the platform services.

The key interest of most platform vendors aims at the provision of efficient services for the handling of general-purpose "objects" (or "documents"). Thus, we shall refer to these frameworks as "general object frameworks". As general object frameworks obviously play an important role in the provision of effective CACE-environments, we shall shortly summarize two of the more recent developments in this field: CommonPoint and OpenStep.

CommonPoint

CommonPoint [8] is a framework developed by Taligent, a joint venture of Apple, IBM and HP. It offers about 100 framework modules covering system and application aspects. It only needs the microkernel of an operating system as a basis. Within CommonPoint, the concept of a file has been replaced by the concept of a general object, which is an interlinkable, recursively constructable (set of) component(s).

OpenStep

OpenStep [9] has been developed by Next on the basis of Nextstep operating system technology and is a.o. supported by Sun and DEC. Microsoft's operating systems Windows NT and Windows 95, which have been developed with the assistance of Next, share many features with the OpenStep technology.

OpenStep consists of three main parts: the application kit, the foundation kit and the enterprise objects framework. The core of OpenStep is defined by a "Portable Distributed Object Model" (PDO) realised in Objective C. PDO-services include storage management with garbage collection, persistency mechanisms, transparent distribution as well as support for CORBA and Microsoft's OLE/COM-Model.

While CommonPoint is new, OpenStep is in industrial use.

2.4.2 Engineering frameworks

The efficient handling of structured, interrelated objects and documents is a basic function also within CACE-environments and hence should be inherited from modern object frameworks. However, management of complex engineering product data and tools places additional high demands on a framework. Moreover, the user interface imposes specific requirements, as the processing of engineering information usually requires high performance graphical interaction in two and three dimensions. Thus, engineering frameworks have been built, which address many of these requirements by offering a computer-based infrastructure tuned to these specific requirements [10].

40

In the historical perspective, engineering frameworks preceded document or general object frameworks. This has been caused by the fact that the poverty of former operating systems was much more serious for highly demanding engineering applications than for office tools, which could also be built on much simpler infrastructures. Here, microelectronic applications and framework R&D efforts were the initiators and driving force of the framework wave in the second half of the eighties (see Barnes [11] for an overview). At the beginning of the nineties the first framework-based engineering environments became available as products by companies as Mentor, Cadence and Viewlogic.

In Europe, a considerable amount of research on engineering frameworks and framework-based environments has taken place in the context of ESPRIT. As a major example, the JESSI Common Frame Project [12] shall be mentioned here. This project also started in the context of microelectronics, but targeted at computer aided engineering or "CAx"-environments. Services for the management of CE-Teams and CACE-Tools were included. Figure 3 shows the conceptual view of the JCF-standard run-time system.

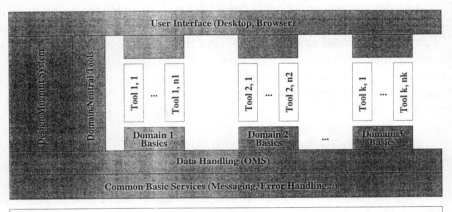

Figure 3   *JESSI Common Framework JCF Conceptual Architecture*

Within the companion projects CONSENS [7] and ATLAS [13], JCF has been successfully applied to various Concurrent Engineering problems. Commercial versions have been derived. Within the project ADVANCE [3] additional common basic CACE-services are under development. Further literature on engineering frameworks is found in [14, 15].

3 CE-MIDDLEWARE

As pointed out in section 2.3., "middleware" between applications and platform services is needed in order to extend the basic platform services towards a powerful support layer for the efficient handling of application data, processes and tools. In the following sections we give examples for middleware services, which are of particular importance for the support of Concurrent Engineering.

3.1 Network and Interoperability Services

The availability of efficient local and wide area network services is mandatory for the construction of modern CACE-environments. Thereby the Internet with its TCP/IP protocol has become the basic international network standard providing interconnectivity on the global level (cf. section 2.3).

On top of the Internet, but not restricted to it, the "World Wide Web" (WWW) [16] has become of utmost importance, especially for the support of geographically distributed engineering teams. Conceived by CERN for easing communication between geographically distributed nuclear physicists, the WWW has been adopted throughout the scientific community. The WWW provides a hypertext-model, a Hypertext Markup Language HTML and a Hypertext Transfer Protocol HTTP by means of which globally distributed documents can be constructed and retrieved.

Though there are practical limits to complexity and performance, the possibility to include multimedia-information and to embed external tools, makes the web suitable for effectively dealing with basic engineering information across the net. Besides the limits with respect to handling complex engineering information, the main shortcomings of the WWW currently are a lack of security and effective navigation and retrieval aids for first-time visited information spaces.

As pointed out in section 2.3, one major task consist in providing better integration between global and local CACE-environments. While at present the WWW undoubtedly is the defacto-standard for achieving a basic level of interoperability on the global level, other standards dominate on the enterprise and the local system levels. Here the Common Object Request Broker Architecture CORBA [17] seems to play an increasingly important role. In [18] an approach is described, of how to integrate WWW and CORBA-based environments via the Common Gateway Interface of HTTP and the Web*-package developed at CERC.

While the services mentioned above supply interoperability on a rather basic level, the recent Java programming language [19] is designed to provide a high-level easy-to-use programming environment for distributed applications. Being architecture-neutral, Java seems ideal for diverse environments such as the Internet. In particular, Java has an extensive library of routines coping with TCP/IP protocols like HTTP. Software written in Java can easily be retrieved from the net and brought to execution on the local networked computer in real time. In future, this may lead to lean local CE-environments, which obtain their power from diverse remote application servers.

3.2 Data and Knowledge Management Services

3.2.1 Database federation

One of the most important requirements for CE-teams is to have access to all data relevant for the given CE-project in an efficient and controlled way. Though many sophisticated database systems are in existence today, the main problem found in industrial environments is the large number of different data storage and management systems, which are often isolated and based on different data models and schemes. Though most data are principally available, they are not accessible in an efficient and controlled way. Cumbersome access procedures, uncontrolled data redundancy, inconsistencies and errors are the well-known consequences in such a situation.

Though a principle solution would be the replacement of the different database systems by a single uniform solution, in most cases this is impractical, as the database systems and their associated applications are the nervous system of the company which must not be disrupted. Thus, a solution is required, which on the one hand enables the continuous access to existing data bases and their applications, while on the other hand addresses the requirements of CE.

Such a solution is provided by the Open Database Middleware System OpenDM [20] (see Figure 4).

Accordingly, OpenDM provides a layer, which federates existing database systems, such that existing local applications (here: A1 and Ax) can further run on their DBS without any recoding, while adapted and newly written global applications may access multiple databases via OpenDM in a uniform and controlled way.

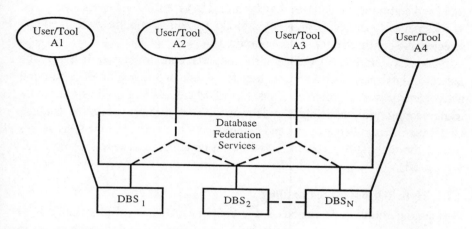

Figure 4   *General OpenDM System Model*

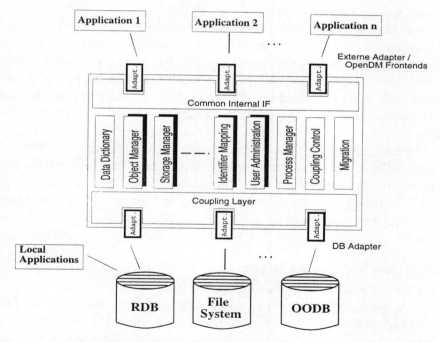

Figure 5   *Overview of Internal OpenDM Architecture*

The principal architecture of OpenDM is shown in Figure 5. Accordingly, tools and component databases may be linked to OpenDM via special adapters. The kernel of OpenDM is given by a toolkit providing modules which can be selected and configured for several purposes.

Besides federation, OpenDM also supports data migration between the federated databases; such a smooth transfer of data and reduction of underlying heterogeneity is also possible. Thus, OpenDM can also be used as a database reengineering tool, assisting the introduction of a more uniform database landscape in large departments and enterprises. In summary, OpenDM solves part of the integration problem between the organizational levels 2 and 3 (cf. section 2.3).

3.2.2 Hybrid knowledge handling

Besides the coupling of databases, the coupling of data and knowledge is of increasing importance. In particular, in CE it is often required to make access to data dependent on constraints and rules and to provide rule-based advice to the engineer.

A viable approach to this problem is described in [21]. Here, a hybrid knowledge description language "HYKL" is used to enable an integrated representation of rule-based knowledge and of complex structured data expressed by means of "frames". Both knowledge and data can be stored in an active object-oriented database, allowing efficient integrated access and knowledge evaluation. A toolkit implementation is available under the name IFS.

In the early phases of the CE-process knowledge is often still uncertain and vague. Thus, it is necessary to also integrate uncertainty information and mechanisms for handling uncertain information in knowledge processing. In [21] it is shown how the HYKL approach can be extended with adequate uncertainty handling schemes.

3.3 Process Management Services

In a CE-environment process control plays a key role [22]. Several teams working concurrently on complex design tasks and doing this potentially in a dislocated manner impose rather complicated problems in controlling this process. Two main areas have to be addressed in order to solve this problem:

- Processes have to be modelled

- Processes have to be enforced and controlled.

Various approaches are known to model complex processes. Here only one of them shall be discussed shortly: Predicate/Transition Nets (Pr/T-Nets). Pr/T-Nets are higher order Petri Nets. The first difference between ordinary Petri Nets and Pr/T-Nets is that in Pr/T-Nets tokens are individuals while in ordinary nets only their count on places is of interest. In Pr/T-Nets a token is an instantiated object of a certain type Assuming an underlying *Petri Net graph* $PG = (P,T,F,B)$, P finite set of *"places"*, T finite set of *"transitions"*, $P \cap T = \emptyset$, $F \subseteq \{(t,p) \mid t \in T, p \in P\}$, $B \subseteq \{(p,t) \mid p \in P, t \in T\}$, a *global marking M* is defined as a bijective mapping from the set of places P onto a partition of the set of instantiated tokens:

$$M : P \to \{K_i \subseteq K \mid K = set\ of\ instantiated\ tokens, K_i \cap K_j = \emptyset\ if\ i \neq j,\ \cup K_i = K\}.$$

This means that tokens are instantiated only as part of the marking of a place and each instantiated token marks only one single place. Input arcs of a transition $t \in T$ are labelled with typed variables. A transition is fireable only if there is a valid interpretation of the set of these typed variables using currently instantiated tokens in the respective places. Equally named variables attached to input arcs of one transition have to be substituted by the same values for an interpretation to be valid. Transitions have attached a predicate and a token mapping. A transition t fires under a specific interpretation only if its predicate is true for this interpretation. So looking for a valid interpretation may be interpreted as looking for sufficient syntactically correct data, while testing the predicate means testing whether semantic restrictions are met. If a transition fires, it destroys the input tokens that constitute the interpretation it is reacting on and calculates (i.e. instantiates) tokens on its output places. By labelling output arcs with typed variables, values can be routed individually to token instantiations.

Example:

Assume the type of all variables in Figure 6a and b to be integer. There are two valid interpretations: $(x \to 4, y \to 5)$ and $(x \to 8, y \to 7)$ but only the first one is accepted by transition t for firing, as the transition's predicate requires $x < y$. As a result of the firing, the marking shown in Figure 6b is obtained.

46

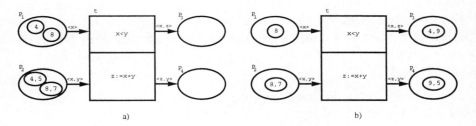

Figure 6   *A Pr/T-Net Transition Before and After Firing*

For complex systems to be modelled in process control the corresponding Predicate/Transition Nets also become very complex and unwieldy. To overcome this problem two concepts of hierarchy have been introduced: hierarchical transitions comparable to subroutine calls of a programming language (the firing of the transition at the higher hierarchical level terminates when the subnet firing terminates) and hierarchical places comparable to the activation and deactivation of processes (the subnet firing terminates when the last token is removed from the corresponding place at the higher hierarchical level). The formal semantics of hierarchical transitions have been adopted from Structured Nets [23] while the semantics of hierarchical places are similar to those of StateCharts [24]. Furthermore recursion can be introduced as well, using these hierarchy concepts [25]. A process control system based on this model will be described in section 4.4.

3.4 User Interface Support Services

A CACE-environment has to serve a heterogeneous community of users. As a consequence, intuitive user interfaces are required, which can be easily adapted to the respective personal style and environment. During the recent years powerful user interface toolkits have come into existence, which support the creation of graphical interfaces and applications [26]. Thereby the "desktop" together with "WIMP"-devices (WIMP stands for windows, icons, menus and pointing devices) have been the driving metaphors.

With the advent of more powerful hardware devices, new multimedia-assisted interaction techniques become feasible and allow the realization of new interaction paradigms and metaphors. Thus, gesture-based interfaces [27] enable real-time recognition and interpretation of gestures, sketches and free-hand drawings - i.e. the support of communications styles typical for the early phases in Concurrent Engineering projects. Virtual reality techniques [28, 29] provide powerful infrastructures for model-based visualization and direct manipulation of engineering artifacts, aiding the early inspection and tele-manipulation of complex engineering products by experts and non-experts in mixed, distributed CE-teams in a natural way. These techniques are only beginning to emerge. Profound changes are directly ahead of us.

4 DOMAIN NEUTRAL CE-APPLICATIONS

While CE-middleware services primarily address support-users (cf. section 2.1), CE-applications are directed at the end-users of CE-enabling technology. Thereby domain neutral applications are those services which are largely independent of a specific application domain, but useful in many different CE-contexts. Their final configuration, however, often is fine-tuned to domain-and/or customer-specific requirements. Below, we shall introduce prominent examples for such domain-neutral CE-applications.

4.1 Global Engineering Networks

As pointed out in sections 2.1 and 2.3, global collaboration and access to global engineering resources is becoming increasingly important for the efficient, timely provision of engineering services and products. While network and interoperability services as described in 3.1 provide a general basis, specific services are required for the effective support of complex engineering tasks. In particular services are required, which support the early phases in product design. Here, it is important to provide access to the large global fund of partial solutions, solution patterns or solution knowledge, which by reuse can prevent an expensive, time-consuming reinvention of the "wheel" or of parts thereof.

For this end, appropriate "global engineering networks" are required, which act as an electronic market place for engineering services and which provide capabilities for efficiently finding and retrieving complex engineering information across global communication infrastructures.

The global engineering network GEN is a collaborative European approach to the solution of this problem. GEN is designed to provide an infrastructure for providers, brokers and customers of engineering services including tele-consultance. While demonstration prototypes have been available since late 1994, the realization of GEN is under way in several projects [30].

4.2 Product Information Management Systems

While databases and database federation systems (cf. section 3.) provide a general infrastructure for storing and retrieving information, the intricate nature of Concurrent Engineering data evolving in course of a Concurrent Engineering process requires a much more sophisticated management support structure addressing at least five different management dimensions, namely [31]: "versions", "views", "variants", "status" and "hierarchy". While "versions" allow for the management of consecutive snapshots during of an evolving engineering object, "views" provide the means for holding related concurrent descriptions of the same object as relevant to different experts in a CE-team and/or different expert tools. "Variants" support the management of different development branches with different optimization goals. "Status" provides a means of expressing the level of quality and completeness of the information at hand. Finally, "hierarchy" allows the break-down of complex engineering data into components and sub-components. It is clear, that all five dimensions are needed for the managing of complex engineering data, whereby the "view"-dimension is of special importance for Concurrent Engineering, as it directly maps to the multidisciplinary approach characteristic for CE.

If we consider the relation between engineering data and processes more closely, it turns out that the five dimensions on the data side may be mapped to five "dimensions" on the process side, namely: "iterations", "design steps", "development of product families", "approval and releasing" and "sub-projects".

The integrated modelling of these five dimensions and their interrelations is a complex problem, which only has been partially solved so far. Moreover, requirements turn out to be quite different for the early design phases, where high flexibility and performance are of key importance, and for the late phases, where controlled formal access is a main issue. Consequently, available product information management systems only address a subset of the problem focusing at a specific engineering phase. An overview of commercial systems is found in [32, 33].

4.3 Workflow Support Systems

The model of Pr/T-nets presented in section 3.4 has been used for process control in a system called DECOR [34]. Figure 7 shows a Pr/T-Net as used by DECOR to model a design process.

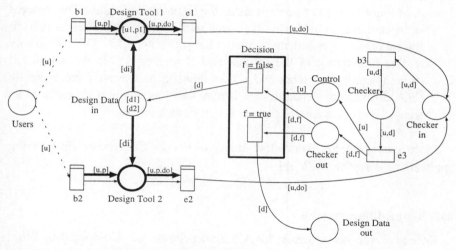

Figure 7   *A Workflow as Modelled by DECOR*

One can see that tools and subflows that are further structured are modelled as hierarchical tool places (*Design_Tool_1, Design_Tool_2*). Tools where no additional details are needed are represented by elementary places (*Checker*). The start and stop of the tools or subflows are represented by transitions (*b1, e1, b2, e2, b3, e3*). Input/output dependencies of tools are represented by data places *(Design Data in, Design Data out, Checker in, Checker out)* that are connected to the respective places or transitions. Control flow places (*Users, Control*) together with transitions (*Decision*) allow the description of design strategies.

DECOR consists of the following components, closely integrated into a CE-environment:

- A Net Editor is used to graphically specify the intended process using Pr/T-Nets. It is used only for set-up and configuration of the Execution Monitor.
- An Execution Monitor provides on-line monitoring of a specified design process. To do this it allows the activation of process actions by firing the associated transactions. DECOR supports concurrent work of a group of users. So only one instance of this process is active at each time so the proper synchronization of concurrent actions can be guaranteed also in a multi-user environment.
- A Monitor User Interface visualizes the actual design status. This process is instantiated individually for each user. This allows each user to communicate independently with the Execution Monitor. DECOR allows on-line monitoring of the process and at the same time its control. It supports concurrent work and allows flows to be executed automatically or in an interactive manner. So, it may serve as a typical example of an advanced process control system of today's technology.

More examples for work flow management in Concurrent Engineering applications are given in [7, 35].

4.4 Joint Editing Systems

Recently, various systems for Computer-Supported Co-operative Work (CSCW) to support joint editing have been reported. Editing in this context includes all kinds of editing: texts, general graphics or even dedicated graphical formats for specific applications. Due to [36] there is a classification of such systems with respect to spatial and temporal distribution. A classical meeting is an example for a synchronous and local action, while a telephone may serve as an example for a synchronous and distributed approach. Asynchronous systems may also be local, e.g. a black board, or distributed like e-mail. Most joint editing systems are synchronous and distributed. *GROVE* [36] may serve as an example of a tool for a distributed editing of text to be used in a conference environment to serve as virtual scratchpad. Such systems usually follow the WYSIWIS-principle (What You See Is What I See) i.e. all participants see the same representation of the jointly edited document down to a single symbol granularity. To avoid access conflicts during concurrent editing, read and write access rights for text segments may be defined. Most graphics editors are bitmap-oriented. Here each modified pixel immediately becomes visible to all users. Of course this means a relatively heavy load on the communication network as all these pixels have to be transmitted. To avoid this the graphics editor *Ensemble* [37] follows an object-oriented approach. Here graphical objects like lines or circles are implicitly reserved for a user whenever they are selected. As a graphical object has to be selected in any case to allow any modification, an access conflict due to various users with concurrent access intentions is avoided. Ensemble works based on the file system. A single process channels concurrent access attempts. Other systems use a client/server architecture. Here the document to be edited is managed by a central server that co-ordinates the distributed applications of the various clients. Fully distributed approaches like GROVE allow each user to keep an individual local copy of the document. All modifications are broadcasted to all other users. On one hand because of this the potential bottleneck of the single server is avoided; on the other hand, however, non neglectable effort has to be spent to ensure coherence of the various copies. In [36] a special coherence algorithm for GROVE is described.

5 DOMAIN-SPECIFIC CE-APPLICATIONS

5.1 Overview

Within Concurrent Engineering various application domains have to be addressed concurrently. These domains have to be supported at the same time and their interdependencies have to be tracked. The design process is guided by different objective functions. These are all "Design for X" objectives, which are discussed in literature. Examples are Design for Manufacturability, Design for Maintainability, Design for Testability and Design for Disposability. In this section a special case of these objective functions will be discussed shortly: Design for Electromagnetic Compatibility.

An important characteristic of CE is the holistic approach it adopts to cover the entire engineering life cycle. System design therefore is an important aspect of CE. In most cases an entire system consists of components from various domains, like mechanics, hydrodynamics, thermodynamics, electronic and some software running on the digital part of the electronic components. A digitally controlled car engine may serve as a typical example where a system consisting of mechanical components, analogue electronics, and digital electronics running under the control of a sophisticated software have to be combined to perform an operation of high thermodynamical and mechanical complexity. Consequently in system design all these aspects have to be considered. Looking at the entire system it can be assumed, that during the design process at the highest level a separation into the different domains of engineering takes place. This partition is a highly complex action, up to now nearly completely carried out by human decisions. CE as an organizational discipline plays an important role in this context as is requires all aspects influenced by such a partition to be involved in this partitioning process. Once a partition has been decided upon and the interfaces have been defined the further design can be carried out within the specific engineering domains. There are many degrees of freedom in this partitioning decisions. The design of a digital system may serve as an example. A priori a solution providing the requested instruction set directly by hardwired hardware is as correct and as obvious as a solution providing the requested instruction set by a piece of software running on a general purpose processor (which is available as a piece of hardware). And within the bandwidth spanned by these two extremes a variety of potential, correct and valuable solutions may exist. Only after one of these solutions may have been selected is the specification of the electronic component to be designed (if not already existing) obtained.

5.2 Goal Oriented Design - Electromagnetic Compatibility

5.2.1 Introduction

Whenever electronic systems have to be designed not only the functionality is of interest but also various physical side effects. One of these effects is called Electromagnetic Compatibility (EMC). EMC-effects like reflection, crosstalk, current spikes, delta-I-noise, electrostatic discharge and radiation may cause malfunctions of the electronic system to be designed or on other systems, even biological ones ("electrosmog"). These interferences are not separately occurring phenomena but have to be considered as inherent properties of electronic systems. As a consequence such systems have to be designed with EMC in mind from the very beginning, a typical point of view of CE. National and international regulations have been implemented to govern this area, e.g. European Community Council Directive 89/336/EEC. Concurrent Engineering enables a holistic approach: EMC-adequate design and analysis methods to the various design stages are applied as early as possible.

5.2.2 The EMC-Workbench: a CE solution of EMC related problems

C-LAB Paderborn has developed the so called EMC-Workbench [38] as an integrated design system for printed circuit design under EMC constraints. The EMC-Workbench is most adequately used as integral part of a PCB design system. This shows the CE approach of this solution. EMC now is no longer a special system feature that is considered independently in a later design step but from the very beginning of physical design. The EMC-Workbench consists of a set of tools for the analysis of EMC-problems on printed circuit boards. These tools are integrated deeply to make sure that all interdependencies are supported properly. The entire workbench is coupled to arbitrary design environments via well defined interfaces.

Some functionalities offered by the EMC-Workbench are listed below:

- Placement analysis (MANDI):
  MANDI enables a fast pre-analysis of the component placement with regard to expected net-lengths and corresponding signal-delays.   In addition a pre-analysis of reflection effects is offered.
- Selective layout data extraction extraction and rule based layout analysis (LDE). The task of the Layout Data Extractor (LDE) is to provide geometric relations between all objects of a printed circuit board layout. As part of this Layout Data Analysis (LDA) is performed, which results in a pre-analysis with regard to reflection and crosstalk effects.
- Simple combined schematic/layout editor for interactive corrections (SCALOR). This editor combines the most important schematic and layout editor functionalities.
- Simulation of reflection and crosstalk effects (FREACS). The simulator FREACS is a highly optimized tool for the efficient analysis of the transmission line behaviour of complex circuits and systems on printed circuit boards.
- EMC macromodel library. To reduce the numerical effort to calculate the non-linear behaviour of terminations of transmission lines appropriate macro models are offered by the EMC-Workbench.

54

- Simulation of radiation and irradiation effects (COMORAN). COMORAN allows calculation of the electromagnetic behaviour of three-dimensional transmission line structures. By application of this tool the engineer is enabled to check for violations of the legal requirements (EMI/RFI limits) without prototyping the actual printed circuit board.

Figure 8 shows the structure and the general design flow of the EMC-Workbench. The workbench layer with its design flow management guides the user, which results in an easy handling of all the highly sophisticated tools integrated in the EMC-Workbench. As an important feature it prevents an inappropriate or even incorrect usage.

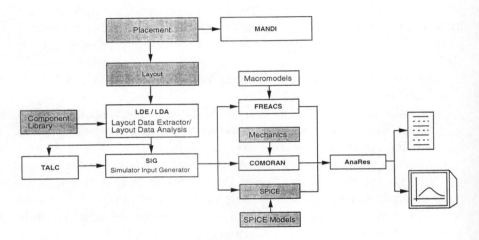

Figure 8   *Structure of the EMC-Workbench*

5.3 Integrated System Design - Mechatronics

5.3.1 Introduction

Since a couple of years the integration of mechanical, electronic and information processing components into so-called mechatronic systems is gaining more and more importance [39]. There is a well proven methodology in the design of mechatronic systems. It is formed by the steps: modelling, analysis and synthesis. All these steps can be supported by computer-based tools. Modelling means representation of the dynamic behaviour of a technical system by means of differential equations or algebraic ones (or a mixture of both). As mechatronic systems consist of components from different engineering domains (mechanics, hydraulics, electronics,..) a more general modelling technique seems to be appropriate, however. The system being modelled and its dynamic behaviour have to be analyzed in detail. Here the identification and separation of the various streams: material, energy and information are of major importance. Synthesis means to build a control system that influences the dynamics of the system to be engineered in the desired way. For the design of linear controllers there are available commercial tools like MATLAB or MATRIX$_X$. Non-linear controllers are more complicated to design.

5.3.2 Modelling of mechatronic systems

For the information processing part of a mechatronic systems the well accepted modelling schemes over various levels of abstraction can be adapted directly. Digital hardware usually is modelled using the following levels of abstraction: At the lowest level, the Electrical Level, the modelling concept is given by differential equations. If the electrical components are abstracted to ideal switches, the Switch Level is obtained. Restricting all possible transistor circuits to logical gates the Gate Level is obtained. In the next step of abstraction a specific mode of operation is assumed (register transfers). As a consequence, this level of abstraction is called Register Transfer Level. At the Algorithmic Level the imperative point of view of the controller is specified. Finally at the System Level it is abstracted even from the algorithmic implementation of the system's behaviour. Only the actors constituting the entire system, the services offered by them (methods in the sense of object-oriented programming) and the interdependent requests for such services are modelled.

This system of abstraction levels can be generalized to entire mechatronic systems rather easily [40]: The System Level now means that the interaction of several subsystems, independently of their nature, is described. The exchanged data now may be elements of the real world, energy, material or information. At the Algorithmic Level an algorithmic solution for the subproblems specified at System Level has to be provided. The only difference to the Algorithmic Level in hardware design is that here the algorithms include actions of the real world, e.g. "transport a piece of material from a to b". The analogy of the Register Transfer Level now is the Functional Block Level. That means that the functional blocks (mechanical subsystems, machines) and their interconnections are identified. Like in hardware design one important aspect is the separation of operational elements from control. The models at lower levels of abstraction are completely contained in the modelling framework of the engineering domains involved. Therefore they need not to be rephrased here.

5.3.3 Analysis of mechatronic systems

In order to obtain a globally optimal system following the approach of Concurrent Engineering it is necessary to specify the entire system and to derive from this specification a partition into hardware, software and non information processing components. The mechanical engineer has to question whether he may decouple energy and information streams (e.g. "fly by wire") or where he may replace sophisticated mechanical devices by simpler ones that are controlled by electronic controllers. The electronic engineer has to analyze the information streams, to identify the algorithms to be executed and to locate them on processing elements. Finally he has to implement these local algorithms in a combination of hardware and software and to install the communication links between the processing elements. Until today the partitioning process has to be performed manually to a large extent. The necessary empirical data can be provided by system simulation. Due to the heterogeneous nature of mechatronic systems, monolithic simulation systems are not appropriate for this task. A solution may be given by multi-simulator systems (simulation backplanes) [41]. In such approaches dedicated simulators can handle subsystems efficiently while only the interaction has to be handled by the backplane. There are three main problems to be solved for this approach: data exchange between the simulators, synchronization of various simulators, homogenization of the user interface.

The synchronization problem is the central one as the other problems can be handled by domain-neutral services of a CACE-environment. There exist two main solutions:
- the (oversynchronizing, pessimistic) supervisor approach,
- the (undersynchronizing, optimistic) Time Warp method.

The supervisor approach at each time allows only this single simulator to proceed which schedules his next event at the closest future point of time of all simulators involved. In the Time Warp method [42] it is optimistically assumed that all involved simulators can run completely independently. If a simulator intends to send into the local past of another simulator, the receiving simulator has to be rolled back to this past point of time. As it itself may have sent messages in the meantime, these messages have to be annihilated using "antimessages" which may cause additional simulators to be rolled back. Of course simulation is not the only tool applicable to system analysis. Static methods are available as well. An example of such a method that is well understood and well supported commercially is performance analysis. Another example is testability analysis. Analysis methods inherited from timing analysis within electronic design can be applied to identify critical paths in concurrent algorithms and to estimate the timing requirements on these critical paths. Similar methods are applied in real time software, called "Schedulability Analysis" in this domain. The most important analysis methods in mechatronics, however, are such ones that help to determine the partitioning of the system into parts for the various engineering domains. It may be assumed that there exists an obvious partitioning into the information processing (IT) part and the environment of it. One natural approach is to identify potential information streams and to investigate whether a mechanical subsystem becomes significantly simpler if it can be reduced to energy transport. The IT subsystem being identified and its interfaces to the environment being defined, this subsystem itself has to be analyzed and partitioned further. A promising approach is to organize the IT part into a system of communicating agents.

5.3.4 Synthesis of mechatronic systems

For the IT subsystem during the analysis step, a set of agents have been identified. Now their internal implementation has to be designed. Both the proper handling of the external interfaces and the internal processes have to be considered. In general a local processing node can be implemented by programming a processor or by designing dedicated hardware. An interesting compromise is given by configurations consisting of a programmable processor and one or more dedicated coprocessors. This approach is called HW/SW Codesign.

Automated synthesis activities are not restricted to the IT part of a system. In analogy to High-Level Synthesis used for digital hardware, in general systems the transformation from the Algorithmic Level to the Functional Block Level can be performed by the following steps [39]:

1. Identification of the functional blocks (machines).

2. Transformation of complex tasks into scheduled sequences of such elementary ones that can be solved by the selected machines directly.
3. Identification of data to be exchanged between the functional blocks (material, energy, information).
4. Identification of where to store this data (material, energy, information).
5. Introduction of connecting structures. For the exchange of material between different machines transport facilities have to be introduced. (Electrical) energy and information can be transported using wires and busses.
6. Identification of the final control structure, i.e. the control part of the algorithm has to be implemented using a controller.

This short discussion shows that methods known from the design of digital HW can be extended to general systems. It is essential, however, that all relevant interdependencies between the various design activities carried out concurrently are continuously tracked by a CE-environment.

6 REFERENCES

[1]   R. Rosenblatt, G. Watson (ed): "Concurrent Engineering", Special Report IEEE Spectrum, Vol. 28, No. 7, July 1991, pp. 22-37

[2]   F. J. Rammig, B. Steinmüller: "Frameworks und Entwurfsumgebungen", Informatik Spektrum 15/92, pp. 33-36

[3]   J. Kaiser: "ADVANCE- Advanced Information Technology for Concurrent Engineering", Int. Workshop Conc./Sim. Engineering Frameworks and Applications, Lisbon, April 1995

[4]   C. Chapel, I. Stevenson: "Concurrent Engineering: The Market Opportunity", Technical Report, Ovum Ltd. September 1992

[5]   G. Almasi et al.: "Functional Specifications for Collaborative Services", CERC Technical Report Series, CERC-TR-TM-94-002, Apr. 1994

[6]   J. Strauß: "CACE System Requirements", Cadlab Report 11/94, C-LAB, D-33094 Paderborn, Germany

[7]   J. Bullinger, J. Warschat (ed): "Concurrent Simultaneous Engineering Systems", Springer ISBN 3-540-76003-2, 1996

[8]   W. Myers: "Taligent's CommonPoint: The Promise of Objects", IEEE Computer, p.78ff, March 1995

[9]   OpenStep Specification; Next Computer Inc., Redwood City, Cal, USA, October 1994

[10] F. J. Rammig, B. Steinmüller: "From Design Environments To Computer Aided Engineering: An Evolutionary Approach", CEEDA'94, Bournemouth, April 1994, pp. 597-602

[11] T. J. Barnes, D. Harrison, A. R. Newton, R. L. Spickelmier: "Electronic CAD Frameworks", Kluwer Academic Publishers, 1992

[12] B. Steinmüller: "JESSI-Common-Framework JCF - An Open Framework for Integrated CAx-Environments", Proc. 5th Int. Conf. HCI, Vol. 19A, p. 337ff, Orlando, Fl, August 1993

[13] Synthetic Overview Report ESPRIT 7280 - ATLAS Architecture, methodology and Tools for computer integrated Large Scale engineering, January 1996

[14] F. J. Rammig, F. R. Wagner (ed): "Electronic Design Automation Frameworks", Proc. 4th International IFIP WG 10.5 Working Conference, Chapman & Hall, London 1995

[15] P. van der Wolf: "CAD Frameworks, Principles and Architecture", Kluwer Academic Publishers, Boston/Dordrecht/London 1994

[16] T. J. Berners-Lee, R. Cailliau, J. F. Groff, Pollermann, CERN: "World-Wide Web: The Information Universe", Electr. Networking: Research, Appl. and Policy, Vol 2, No 1, pp.52-58, spring 1992, Meckler Publ., Westport, USA

[17] Object Management Group Inc.: "The Common Object Request Broker: Architecture and Specification", Rev. 2.0, Framingham, MA, USA, 1995

[18] G. Almasi, V. Jagannathan: "Integrating the WWW and CORBA-Based Environments", CERC Technical Report Series, TR-RN-95-005, Morgantown, WV, USA, September 1995

[19] Sun Microsystems: "The Java™ Language: A White Paper", Mountain View, Cal., USA, 1995

[20] R. Böttger, Y. Engel, G. Kachel, S. Kolmschlag, D. Nolte, E. Radeke: "Enhancing the Data Openess of Frameworks by Database Federation Services", Conf. on Electronic Design, Automation Frameworks (EDAF), Grammado 1994

[21] P. Drescher, M. Holena, R. Kruschinski, G. Laufkötter: "Integrating Frames, Rules and Uncertainty in a Database-Coupled Knowledge-Representation System", Proc. DEXA 1994, Athens, September 1994, pp. 703 - 712

[22] J. B. Brockmann, S. W. Director: "The Hercules CAD Task Management System", Proceedings ICCAD, 1991

[23] L. A. Cherkasova, V. E. Koov: "Structured Nets", Proc. MFCS'81, Springer LNCS 118, 1981

[24] D. Harel: "StateCharts: A visual formalism for complex systems", Science of Computer Programming 8, 1987

[25] F. J. Rammig: "System Level Design", Mermet (ed), Fundamentals and Standards in Hardware Description Languages, pp. 109-151, Kluwer, 1993

[26] B. A. Myers: "User Interface Software Tools", ACM Transactions on CHI, Vol.2, No.1, pp. 64 - 103, 1995

[27] R. Zhao: "Handsketch-Based Diagram Editing", Teubner-Texte zur Informatik , Stuttgart, Leipzig 1993

60

[28] G. Singh, S. K. Feiner: "Introduction to the Special Issue on Virtual Reality", acm Transactions on Computer-Human Interaction, Vol. 2, No. 3, September 1995, pp. 177-179

[29] L. J. Rosenblum, S. Bryson, S. K. Feiner: "Virtual Reality Unbound", IEEE Computer Society, Vol. 15, No. 5, September 1995, pp. 19-22

[30] 1st European Workshop on Global Engineering Networking, C-LAB and HNI, Paderborn, Germany, February 1996

[31] P. van den Hamer, K. Lepoeter: "Managing Design Data: The Five Dimensions of CAD Frameworks, Configuration Management and Product Data Management", Proceedings of the IEEE, Vol. 84, No. 1, January 1996

[32] E. Miller: "PDM Today", Computer Aided Eng., February 1995, p.32 - 40

[33] W. Collier: "Converging on Product Information Management",Computer Aided Eng., May 1995, pp. 64 - 74

[34] J. Tacken, L. Kleinjohann: "Management of Concurrent Design Processes", CEEDA'94, Bournemouth, April 1994, pp. 43-48

[35] M. Hsu (ed): "Special Issue on Workflow Systems", Bulletin of the Technical Committee on Data Engineering, March 1995, Vol. 18, No. 1, IEEE Computing Society

[36] C. A. Ellis, S. J. Gibbs, G. L. Rein: "Groupware: Some Issues and Experiments", Proc. of the ACM 34 (1), pp. 38 -58, 1991

[37] R. E. Newman-Wolfe, M. L. Webb, M. Montes: "Implicit Locking in the Ensemble Concurrent Object-Oriented Graphics Editor", CSCW92: Proc. Of the Conference on Computer-Supported Co-operative Work, pp. 265-272, November 1992

[38] W. John: "Support of Printed Circuit Board Design by an EMC-Workbench", Proc. of the 10th International Zürich Symposium and Technical Exhibition on Electro-magnetic Compatibility, Zürich, 1993, pp. 185-194

[39] M. Acar, I. Makra, E. Perrey: "Mechatronics, the basis for new Industrial Development", Computational Mechanics Publications, 1994

[40] M. Brielmann, F. J. Rammig: "Evaluating Hardware Design Principles for the Development of Computer Based Systems", Proceedings IEEE Symposium & Workshop on ECBS, 11. - 15.03.1996, Friedrichshafen

[41] M. Niemeyer: "Simulation of Heterogeneous Models With a Simulator Coupling System", Proc. SCS'91 European Simulation Multiconference, June 1991

[42] D. R. Jefferson, H. A. Sowizral: "Fast Concurrent Simulation Using the Time-Warp Mechanism", Proc. SCS Distributed Simulation Conference, 1985

7 ACKNOWLEDGEMENTS

The authors would like to thank the staff of C-LAB for valuable contributions and in particular W. Fox and M. Joosten for the review of this document and many helpful comments.

PART II

PRODUCT DESIGN, DEVELOPMENT AND CONTROL

CHAPTER 3

Designing for the Life Cycle

M. J. Hall

1 INTRODUCTION

The concept of the product life cycle has been with us for some time. However, the importance of focusing product design teams on all aspects of the cycle is increasing.

This increasing focus is driven by rising consumer expectations, shorter life cycles, technology change, increased environmental concerns and a host of other factors. With modern technology the impact of the designer on life cycle parameters is increasing. For example, over 80% of the product cost is committed during the early stages of the design process, (ref. 1); the reliability of the product is virtually decided during design and the ability to re-cycle is largely design dependant. This means that the engineer working in the product creation process has to be aware of, and design for, the whole product life cycle.

The requirement to satisfy the customer throughout the product life cycle puts a large and complex set of demands on the design team. These can only be met, effectively and economically, by a multi-disciplinary team using the right methods and tools.

The product creation process now calls upon the expertise of people from a wide range of business functions. Design is no longer the exclusive preserve of the design engineer, if it ever was. True multi-disciplinary team working is essential but can be stressful and difficult to manage. New methods of project management have to be employed to ensure success. These start with the process of creating the product design.

2 THE PRODUCT CREATION PROCESS

a) The Process

The process of creating new products starts with an understanding of the market place and the needs of the customer. This is the realm of the marketeer on whose skill largely depends the success or failure of the product.

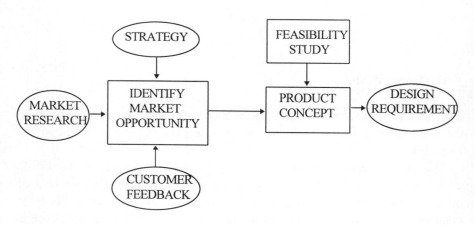

Figure 1   *Design Requirement*

The job of the design team starts with interpreting the marketeer's concepts, or the customer's requirement specifications and turning them into reality (Fig. 1). To be competitive though, the product creation process needs to be ever faster and cheaper, while producing designs that address the needs of the whole life cycle. Concurrent Engineering is a good approach for doing this.

The marketeer or customer's design requirement is turned into a product definition that specifies all of the life cycle parameters. This must be easily understood by all team participants so that plans can be made and design constraints identified.

The various functions work concurrently with effective interchange of information (Fig. 2). The most important aspect of this is teamwork, although good data base access and linked systems help.

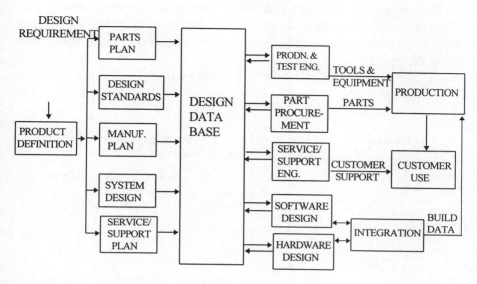

Figure 2   *Typical Concurrent Engineering Process*

b) Balance

The execution of the product creation process involves a lot of compromise. There are four factors to balance when running a product design project and they all interact.

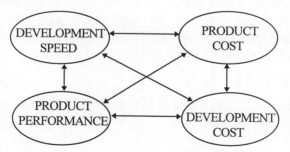

Figure 3   *The Six Trade-offs between Product Development Objectives (ref. 3)*

Product development models, such as a Product Profit model, can be used to achieve the best balance for a product in a particular market.

66

The speed of development is largely driven by the market window (Fig.4). Time to market, particularly in fast moving markets, often makes this the most critical of the four factors. The effect of delayed market entry on revenue can be approximated by:

$$\text{Lost Revenue} = \frac{d(3w-d)}{2w^2} \times 100\%$$

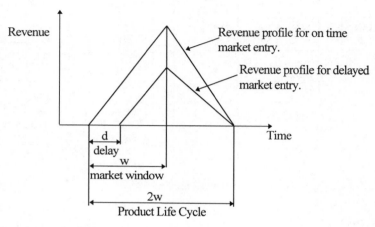

Figure 4   *Revenue Lost from Delayed Market Entry*

The Development Cost is the total cost of creating a new product, putting it into production and setting up customer support (Fig. 5). It should be viewed like any other investment, both in terms of how much should be spent and what level of return makes it feasible. The life time marginal income is the cumulation of sales less cost of sales over the life cycle. Net present value is used because the life time may be several years.

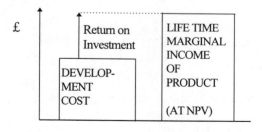

Figure 5   *Product Investment Payback*

The Product Cost is the whole life cycle cost of the product to the customer. This includes the initial purchase price and represents total cost of ownership. The relationship between Product Cost and sales volume comes from market research (Fig. 6). It may, of course, change with time.

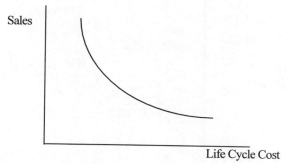

Figure 6   *Sales vs. Life Cycle Cost*

The Product Performance is described by a range of factors which affect the user at different points in the product life cycle (Fig. 7).

In defining the product it is necessary to establish the relative importance of the factors for that market and not under or over specify any of them. This is achieved through the application of market knowledge in a structured way to all aspects of the product life cycle.

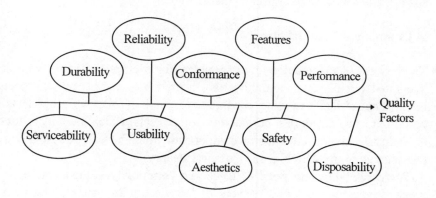

Figure 7   *Quality Factors*

3 THE PRODUCT LIFE CYCLE

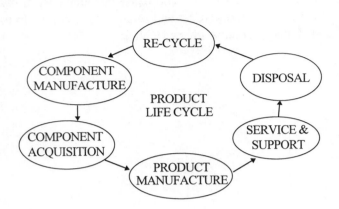

Figure 8   *The Product Life Cycle*

This is the cradle to grave journey that the product is intended to follow. In terms of most manufactured products this starts with mining the ore and finishes in the dustbin, or rather the re-cycling plant.

Throughout the life cycle two measures are of primary importance: costs and cycle times. Costs are incurred from component manufacture through to disposal and most of them are picked up by the customer. Cycle times also occur throughout the life cycle from component lead times through to disposal; again, these have a large impact on the customer.

a) Component Manufacture

The process of manufacturing components (for example, resistors, capacitors, integrated circuits and circuit boards), is normally carried out in specialist manufacturing facilities rather than general assembly plants. Components can be put into two categories: commodity components, which are widely used in the industry and custom or drawn components, which have to be specially made (circuit boards and ASICS come into this category).

The availability and price of electronic commodity components are very much driven by the largest sectors of the electronics market, like personal computers and mobile phones. When designing for the future it is necessary to understand the rate of technical change and project future prices of the components concerned (Fig. 9).

Introduction Year	DRAMS	Relative Cost/MBit
1989	4M	100%
1992	16M	43%
1995	64M	21%
1998	256M	10%
2001	1G	6%
2004	4G	4%
2007	16G	3%

Figure 9   *Example, DRAM Memory Development (projected)*

The unrelenting and increasing pace of change in electronics also means that obsolescence risk can be high. This can cause high costs in the later stages of a product's life cycle.

Custom components require a different approach with close technical links between supplier and user. Designs need to keep pace with the suppliers technical development programme and take full advantage of any new technologies.

b) Component Acquisition

The component acquisition process starts with a demand or order and finishes with the delivery of the material to the shop floor process that requires it. The acquisition process is an important part of the life cycle because it can add up to 20% onto material costs and account for up to two thirds of manufacturing cycle time. The start of the acquisition cycle can be driven by forecast, actual demand, a mixture of both or, if demand is fairly steady, a schedule that is periodically updated. The acquisition process adds cost by employing people and assets and by attracting interest charges on material while it is in the factory. The basic process consists of:-

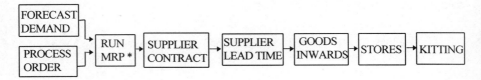

* MRP = Material Requirement Planning, a computerised system for stock control and
  purchase order generation.

Figure 10   *The Material Acquisition Process*

The main cost drivers within this process are the number of part types,
number of suppliers and inventory levels.

In situations where demand is stable it is possible to streamline this process
and adopt just in time (JIT) principles. The supplier base can be reduced and
run on long term price/quality agreements with demand communicated as
regularly updated schedules. Close supplier partnerships can be established so
that delivery of parts can be straight to the shop floor with no storage or
inspection. This greatly reduces costs and cycle times but is not always possible
due to fluctuating demand.

c) Product Manufacture

The manufacturing process for electronic products can be divided into three
stages: component manufacture, circuit board assembly and product assembly.

i) Component manufacture is largely carried out by specialist manufacturers
and is replaced by the component acquisition process.

ii) Circuit board assembly has now become a very automated process that is
very capital intensive, uses hardly any labour and requires good process control
to achieve acceptable quality levels. The lowest costs per component placed are
achieved through economies of scale, and the surface mount lines component
placement rate is the most important factor. The most economic lines operate at
speeds of approximately 65,000 components per hour, but are only suitable for
volume products with good part rationalisation (less than 80 component types

on each). Such lines have a capital cost in excess of £2M with fully integrated inspection and test capability. For lower volumes, more set ups and higher numbers of part types, it is necessary to compromise, which increases the cost per placement.

iii) Final product assembly can benefit from automation but, unlike circuit board assembly, automation at this stage tends to be inflexible and shorter product runs become uneconomic. Usually the best design approach in these circumstances is to design for automated assembly and then do it manually.

Costs and cycle times are lowered by minimising the number of process steps and working at minimal work in progress (WIP) levels. This requires close coupling of process steps which can be achieved by one piece flow lines on volume products. For lower volumes, cell-based manufacturing, which close couples processes that are not continuous, will achieve the same result.

The ideal manufacturing installation produces one product, or a range of very similar products, in volumes that allow the use of state of the art placement machines, operating continuously seven days a week. The finished product is assembled on flow lines directly coupled to circuit board assembly. Components are loaded at one end of the line and boxed product ready to ship comes off at the other. Mobile phones are an example of a product that can be produced in this way.

d) Service & Support

This part of the life cycle may require a full range of support services or it may require very little, depending on complexity. Most products require some form of maintenance which may be just breakdown maintenance on simpler products or extend to preventative maintenance on more complex ones. The three key elements to providing a maintenance service are:

i) Cost - there is a service cost threshold above which the product becomes a throwaway. Increasingly service is sold with the product in the form of an extended warranty making it an important cost to control.

ii) Turnaround time - the time taken for the product to get back into action after a breakdown. Expectations are getting shorter, particularly with products that earn the supplier revenue when used, like phones. Very short service cycle times can be achieved by operating a replacement service rather than repairing the actual product that has broken down, but this increases cost in the form of stocks.

iii) Quality - the quality provided encompasses all aspects of the customers interaction with the service provider and also the technical quality measured by the 'bouncer' rate, which is the number of repaired products that breakdown again within the service warranty period.

Other aspects of post sale support can include training and customer help lines, which are an increasing feature provided with personal computers. With professional products the actual cost of use becomes a critical element in the customers choice of a particular model.

e) Disposal & Re-cycle

End of life disposal of electrical products is the part of the life cycle that is getting increasing amounts of attention as traditional landfill solutions become unavailable. In the UK the waste stream for domestic electrical products has topped one million tons per year and includes 1.6 million microwave ovens, 3 million phones, 1.4 million televisions and 800,000 computers (ref. 7). The method of disposal and hence the cost of disposal is designed in. The design life of products is getting shorter because the pace of technical change is such that consumers will not 'live with' old technology.

The driving factors behind re-cycling are legislation and costs. The legislation will get tougher and the costs will come down.

In order to address the whole life cycle, a range of life cycle parameters need to be considered.

4 LIFE CYCLE PARAMETERS

The main performance parameters of a product through its life cycle can be divided into three groups: costs, cycle times and quality factors. These performance parameters all have drivers within the design process which get committed fairly early. For example, it is typical for 80% of life cycle costs to be committed early in the design process.

a) Costs

Product costs can be divided between factory costs, business margin and post sale costs, which all add up to the total life cycle cost. This is the full cost incurred by the customer in buying, using and finally disposing of the product. The list of costs is not exhaustive but serves as an indication (Fig. 11).

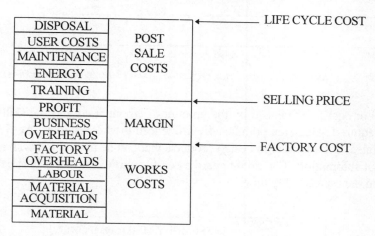

Figure 11 *Life Cycle Cost Stack*

i) Material cost is the actual cost of bought-in parts and in modern electronic products it makes up over 70% of works cost (Industry sources). This makes part selection one of the key drivers in the design process. The cost of commodity components is driven by the high volume users in world markets; everyone else has to watch and predict. It is very important to select components that will fall in price during the manufacturing life of the product if the cost of that product is to be reduced year on year (Fig. 12).

ii) Material acquisition cost is the cost of obtaining the material and delivering it to the point of use. This will include purchasing, material control and store costs. The drivers of material acquisition cost are the number of suppliers (the lower the number the lower the cost), the number of part types (again, lower is cheaper) and the component package technology (surface mount components are easier to handle than leaded). Other factors affecting acquisition cost are volume, volume stability and the locality of suppliers.

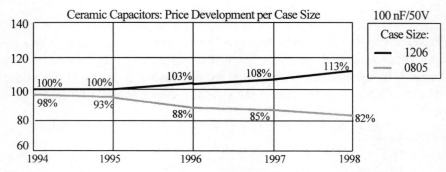

Figure 12   *Example of Forward Price Projection (100nF, 50V, ceramic capacitors)*

iii) Labour cost is influenced by the amount of labour used and the skill level that is required ,which are both heavily influenced by the designer. At Western labour rates, it is essential to design products that can be assembled with a high degree of automation. For circuit board assembly surface mount is best because it has the lowest assembly time.

PROCESS TYPE	ASSEMBLY TIME/COMPONENT
HAND	7 SECS
LEADED AUTO	1.5 SECS
SMT	0.6 SECS
HIGH SPEED SMT	0.12 SECS

Figure 13   *Component Assembly Times for Different Processes*

Also driving labour cost are the numbers of parts, part types and assembly directions which should all be minimised.

iv) Factory overheads are usually dominated by three groups of cost: the site itself (land, buildings), indirect labour and the cost of capital equipment. Site costs are a function of location and size, which are not very influenced by design except for the economics of scale achieved by addressing global markets. Indirect labour is a function of technical complexity and the amount of labour to be managed; these can be minimised by minimising labour content and technical complexity. The cost of capital equipment should not necessarily be minimised as automation reduces labour content, but the amount of capital investment can be reduced by reducing part variety and by optimising product designs to machines. In today's fast moving environment capital assets need to be worked very hard in order that they can be written off quickly and replaced with newer technology.

v) The margin generated to cover business overhead and profit can be influenced by design costs. Minimising overhead creates more profit which allows more re-investment.

Factory + Margin = Selling Price
Cost

Margin is expressed as a percentage of selling price and varies from as low as 15% for high volume products up to as high as 50% for capital goods.

The post sale costs that dominate will vary a lot with different products but all are very influenced by design.

vi) Training - the cost of learning how to use a product is very much a function of complexity and 'design friendliness'. The use of Usability analysis during design can decrease this aspect of cost.

vii) Maintenance - these costs are a function of reliability and cost of repair. Reliability (mean time between failures) is very dependant on component selection and derating (the amount of stress a component is put under during use). The cost of repair is also very dependant on part selection and ease of disassembly.

viii) Energy costs for electronic products are a function of power consumption and consist of battery replacements or mains electricity use. Power consumption is very much a part of the design brief.

ix) User costs represent the actual cost, usually time, of using the product which is very important for professional products that are used in a commercial environment.

x) Disposal costs are now rising as environmental protection laws are passed and it is no longer acceptable to dispose of products in land fill. The cost of disposal can be minimised by designing for easy disassembly, minimising the use of hazardous materials and using raw materials (like certain plastics) that can be reprocessed.

b) Cycle Times

Cycle times affect customers directly in terms of the responsiveness of a company to orders or service requests and indirectly in the form of increased costs if cycle times are long. The main cycles in the life of an electronic product are component acquisition, manufacture, service and disposal.

The component acquisition cycle is the time from a demand request to the component parts being available for use in the factory. This cycle is made up of various scheduling and ordering tasks, dependant on the material control system that is used and the actual part lead times from suppliers. The material control tasks can be speeded up by restricting part variety within products. Supplier lead times are dependant on component choice. If a short acquisition cycle is needed then only short lead time parts should be used in the design.

The production cycle time is dependant on the number of processes that have to be used and the time taken for each process. The circuit board assembly process is dependant on the mix of component technologies and whether the board is single or double sided (Fig. 14)

	TECHNOLOGY	PROCESS STEPS
S1	Single Sided SM Reflow	3
S2	Double Sided SM Reflow	6
S3	Mixed SM/TH SM Single Sided	6
S4	Mixed Assembly SM Double Sided	9

SM = Surface Mount
TH = Through Hole

Figure 14   *Alternative Board Assembly Processes*

Each stage adds to both the cycle time and the defect rate which, subsequently, increases rework.

Each process cycle time has three elements: run time, set time and queue time. The run time is the time taken to complete one assembly and is largely driven by complexity. The set time is the time taken to set up between the production of different product designs, which is driven by the amount of component variety. If all of the products on a line consisted of the same set of components the set time would be minimal. The queue time is the time a batch of product has to wait to get onto a particular process, i.e. it sits in a queue waiting for preceding batches to finish. This is dependant on machine utilisation, so if machines are heavily utilised (90% +) the queues will be long.

The service cycle time (repair of a faulty product in use) will depend on complexity, ease of fault diagnosis and ease of disassembly, all of which are design driven. The principles applied on a production line are similar to those in a repair workshop in that there are a number of stages with run, set and queue times.

The disposal cycle time will depend on ease of disassembly, the use of hazardous components (and their method of safe disposal) and the ease of processing of the raw materials used. An example of an exceptionally long disposal cycle is a nuclear power station, because of the hazardous material used and the fact that disposal was probably not considered during design.

c) Quality Factors

The Quality Factors represent the main operating characteristics through a product's life (Fig. 7). When defining a product the factors should be ranked in order of importance for that product and quantified for the design team. If a factor is not quantified the outcome of the design process will be uncertain and, almost certainly wrong.

i) Performance - the key performance parameters of the product have to be defined and quantified. This should include not only the target value of each parameter but also the limits on its variability (tolerance). Meeting tight tolerances on performance can pose as difficult a design task as achieving the intended level. Because of this it is vital to target levels and tolerances adequate for the market, not too high or too low.

ii) Features - these are the optional extras in the specification that the customer can choose. They have to be specified in terms of performance and cost. Ideally, features should be designed for fitting late in the manufacturing process.

iii) Reliability - the reliability of products is usually expressed as a failure rate or mean time between failures (MTBF). The failure rate of a product is decided by a number of factors. The main ones are the reliability of individual components, the degree of stress that each component is under and the physical environment that the product is used in. Component choice is vital to this in that components should suit the environment and be of similar grade with no weak links. Components should be run below their specified stress levels, and this is called derating. Different types of component will have different parameters that cause stress, for example, with resistors it is power, capacitors voltage and digital I.C's fan out (the number of inputs connected to each output).

iv) Durability - this is really two parameters, the design life of the product and its environmental specification. The design life is the length of time it will last in normal use and may range from a few hours up to thirty years or more across the whole range of markets. Life may also be expressed in terms of operations or other variables. For example, a car's life can be expressed in miles or years. The environmental specification is a description of the physical environment in which the product has to operate in terms of temperature, humidity, vibration and a range of other parameters.

v) Serviceability - the ease with which a faulty product can be repaired will decide its cost of repair and will influence the decision as to whether to repair it or throw it away. The time taken to effect a repair is expressed as the mean time to repair (MTTR) and is influenced by the physical construction of the product. Also important are the cost and availability of spare parts. If a product is designed as a throw away (no repair provision) then serviceability can be taken out of the design brief.

vi) Aesthetics - the basics are physical size, weight and colour. Styling is very important for consumer products as is the ability to change aesthetic parameters cheaply in order to refresh products coming to the end of their market life.

vii) Conformance - there are many standards for products around the world, some influence customer choice and others are a legal requirement. It is vital to list all of the essential and desirable standards for a product in all of the markets in which it is going to be sold. The requirement to conform to standards can have a major impact on design and manufacturing.

viii) Usability - the ease of use of a product, particularly when being used by a customer for the first time is a major factor in consumer choice. Usability is affected by ergonomics, the man machine interface (MMI) and the clarity of instructions. This factor is often tested on consumers in trials and laboratories before product launch.

ix) Safety - the safety requirements of a design can be assessed in terms of the potential impact they can have on risk of injury or fatality.

x) Disposability - if a product is going to be disposed of with no environmental hazard and maximum re-use of its raw materials, it has to be designed that way. The product definition needs to specify the end of life disposal process that the product is being designed for and that process has to comply with environmental regulations in each country. This is particularly difficult with long life products because the regulations can change completely over a period of ten years or more.

In order to effectively design for the life cycle a structured process is required.

5 DESIGNING FOR THE LIFE CYCLE

In order to adequately address the many aspects of designing for the full life cycle, it is essential to consider the actual design or product creation process. This cannot be left to chance and needs to be clearly structured for all of the participants in advance.

There are three distinct phases that need to be addressed in order to arrive at a finished product design (Fig. 15).

Figure 15   *Three Stages of Design*

The earliest stages are the most important, as they are where the key decisions are made and where the bulk of the costs are locked in. Premature progress to implementation, the actual engineering, makes the whole process a lottery and greatly reduces the chances of creating a commercially successful product.

a) Product Definition

A product can be defined in terms of its life cycle parameters. The targeting of life cycle costs starts with price projections for the product in its market place. These will range from its launch price to its price when it is finally withdrawn from the market. Most electronic products fall in price during their market life. Margin levels are set in order to pay off investment in design and plant, cover business overheads and make a satisfactory profit. The projected works cost can then be used as a basis for target works costing. For complex products this target cost can be divided up across parts of the product. As target works costing is essentially a management tool, this division should mirror the division of the design team into working groups in order to provide each group with a clear target for the designer. As it is expected that the price of an electronics product will fall during its life, it is reasonable to target the works cost to fall also.

In addition to targeting the works costs of a product, the post sale costs should also be considered in order to assess the full life cycle cost. The importance of targeting post sale costs will vary depending on the market. Products with a very long life (for example military products) may have post sale costs many times higher than their purchase price, making these costs key to gaining competitive advantage. Consumer products that are covered by extended warranties require the targeting of service costs for the extended warranty period if a profit is to be made on the sale of such warranties. If a product has to contain hazardous materials (for example a nuclear component) the cost of end of life disposal should be targeted as this could easily exceed all of the other costs.

The next part of the product definition is the targeting of cycle times. The four key cycle times are component acquisition, production, field service and disposal. The component acquisition cycle time will depend on a number of market factors. If the customer lead time is short, the purchase of components will be forecast driven. For longer lead times it may be possible to make it order driven, reducing the risk of surplus stock write offs. If a market is very volatile with demand peaking and troughing rapidly, it will be very difficult to run a system based on forecasts without accumulating large component stocks, which is expensive. If the market conditions lead to a situation where the demand is volatile, it is advantageous to target a short cycle time for component acquisition. With stable markets, a longer cycle time is possible. There will be a compromise here between costs and cycle times because shorter acquisition cycles usually lead to higher material costs.

The production cycle time should always be as short as possible in order to minimise stock levels. The main driver of this cycle time is product complexity so this target will depend on the technical part of the definition. As with material acquisition, production may be forecast driven or demand driven according to the market. The trend is to try and be demand driven in order to eliminate costly finished goods stocks, which increases the pressure for shorter cycle times.

The service cycle time target will depend upon the service strategy that is adopted. Products that generate income while operating (for example mobile phones and cash transaction systems) will require very fast service turnround. However, if this is achieved by providing an exchange service, there may not be a requirement for a very short (and expensive) repair cycle.

The targeting of disposal cycle times will depend on the environmental legislation that applies. If the re-cycling of a product can generate income for the re-cycler, the net cost of disposal could be zero. This will make short disposal cycles a priority, because a long cycle means more expense or more long term environmental damage.

The market pressure, generally, on cycle times is for them to get shorter. It is therefore logical to always target shorter cycles from each product generation to the next.

The final element of the Product Definition is the list of Quality Factors which represent all of the technical aspects of the product. To start with they should be ranked in order of importance for each individual product as it is rare for them all to have equal emphasis. Factors should then be quantified in as much detail as their place in the ranking warrants. It is very important to quantify all of the factors that will drive design decisions because leaving them to chance will virtually guarantee a wrong result. The detail should be driven purely by the needs of the market place. Overspecified parameters will lead to excessive costs and underspecified ones to a weakened competitive position.

QFD (Quality Function Deployment) is a very useful tool for matching market need and design parameters.

b) Design Planning

The design planning stage is where all of the key technical decisions are made. The multi-disciplinary team must debate all aspects of the design because they all interact.

The most critical tasks when starting up a new team are to establish its purpose, the process by which it will operate and measures of team progress. Individuals should be brought together who will work well as a team, by determining whether each person has the knowledge, skills and influence to participate effectively.

i) The purpose of the team can be established by defining why it exists. The team should discuss who its customers (internal or external) are and how their needs can be balanced.

ii) The process of operation includes team leadership, decision making and what tools and techniques will be used. It may be appropriate to use a facilitator to manage conflict and encourage participation.

iii) Measures should be agreed that indicate that the team is making progress and whether it has reached success or failure.

The main elements of the design plan are:

- The component set which details the types and ranges of components from which the product will be designed. This will be selected on the basis of technical requirements, component prices with forward price projection over the life of the product, obsolescence risk and lead times. Component rationalisation is a very important driver of both costs and cycle times. It can be targeted as a ratio of part types to parts used. The choice of component technology is also closely linked to the manufacturing process technology that will be used. If existing assets are going to be used in manufacturing, the component technology will have to be matched to them.

- A list of manufacturing processes, including test stages, detailing the technology that will be employed and any new technology that requires development. Each manufacturing process, including those used by vendors to produce custom parts, will have associated with it a set of design constraints or guidelines. It is essential to identify and comply with these in order to achieve the best quality and performance from each process.

- As with manufacturing processes the post sale activities such as service and disposal will also generate design guidelines. It is therefore essential to identify them at this stage.

- The design itself should be outlined in terms of its basic structure. For complex products the system design level should be completed. This allows for the planning of the design at module level.

- Because both time and money will be constrained when creating a new design, it is vital to make maximum use of existing elements from previous designs. They may need to be modified to conform to the latest specification but this can still be the most cost effective solution.

- The design tools that are going to be used should be defined at the outset in order to avoid any interfacing problems. These will normally include circuit layout CAD, mechanical CAD, circuit simulation, test simulation and configuration management. Specialist tools in areas like thermal management, stress analysis and reliability prediction may also be used.

- Design reviews need to be held at appropriate points in the process to ensure that technical risks are acceptable and to check financial parameters. A useful technique for technical review is the so called 'Red Team Review'. This requires a team to be assembled independently of the design team but with similar skills, whose purpose is to check all technical aspects of the design and make comments. The process requires the participation of the original team in order to answer questions and make presentations. This can take several days to complete. The financial review is to audit project costs against the plan and to forecast cost to completion.

Planning the design and then managing its implementation will require an appropriate form of project management if it is to be successfully controlled and co-ordinated. With the use of multi-disciplinary teams this can be a complex task. One planning method particularly suited to this environment is Goal Directed Project Management (GDPM) (ref. 4). This approach actually uses the multi-disciplinary team to formulate the detailed project plan in a way that ensures that everybody buys into it. It also shares out the management task throughout the team by allocating milestones to milestone owners. This is not necessarily done on a skill match basis; for instance, a manufacturing engineer can be a milestone owner for a design task. However, it is important that the milestone owner understands the content of the task and is supported by the rest of the team in its achievement. Timescales are focused onto meeting customer, or market, requirements. The basic planning stages of GDPM are:

- Identify milestone tasks.
- Put required completion dates on milestones, taking account of inter-dependency between them.
- Allocate milestone owners.
- Describe the milestones in detail, dividing into sub-milestones if complex.
- Estimate resources for each milestone.
- Cost the plan and do a feasibility check.

The stages are carried out by a multi-disciplinary team, not in isolation.

c) Implementation

The prospect of designing a product with such a large number of design considerations may seem a very complex task but it can be greatly simplified. This is because many of the life cycle parameters have common drivers making it possible to achieve a good result by applying just a few basic principles.

i) The most important design consideration for an electronic assembly is component selection. The components account for over 70% of works cost, influence all key cycle times and affect most of the quality factors. Component selection should take account of the following:

- The price of the selected components should be projected to fall annually for the duration of the manufacturing run.
- There should be a low obsolescence risk for the duration of the products service life.
- Component reliability should be compatible with that required of the finished product.

- Component lead times should be compatible with the component acquisition cycle time.
- A good match needs to be achieved between components and the target manufacturing process.

It is worth putting a lot of effort into selecting a good component set and then designing within it. Component reliability and therefore product reliability, can be improved by derating components. This is achieved by referring to derating charts which give the percentage electrical stress that can be applied to a component at different operating temperatures and in different environmental conditions. For example a 10V capacitor will be more reliably operated at 8V than at its fully specified 10V. Excessive derating can be expensive and amounts to over-engineering.

ii) Another important design consideration is the relationship between a products performance and its specification. This relationship will affect test yields, test and rework costs and field reliability. The variability of performance parameters against specification is also important, especially on higher volume products.

Ideally, six sigma design principles are applied such that the relationship between the specification and the specification limits is $\pm$ 6 sigma of the distribution achieved in manufacturing (Fig. 16).

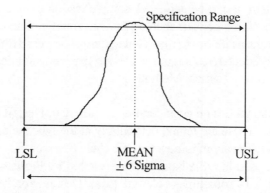

Figure 16   *Specification Range*

In theory such a tight spread would give a defect rate in manufacturing of only 0.002 defects per million (dpm) but in practice a shift in the mean of 1.5 sigma is allowed for to cover sampling errors and time to time variations in the mean (Fig. 17).

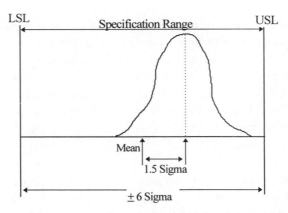

Figure 17   *Specification Range with Shifted Mean*

This level called mean shifted six sigma performance would give a defect rate of 3.4 dpm, which is still very low. Design to these standards eliminates the need to do 100% testing which greatly reduces manufacturing costs, rework and field failures. It is not usually possible to achieve this level on all performance parameters but where it can be achieved, sample testing can be adopted. The achievement of six sigma performance depends as much on the realistic setting of specification limits as it does to the detailed design process. If the variance of a design has to be reduced, techniques involving experimental design will need to be used (for example, Taguchi Methods).

iii) The third important design consideration is the use of Design For Assembly (DFA) techniques. These are aimed particularly at the mechanical aspects of a design and have a wide ranging impact on life cycle parameters.

The techniques used in DFA have been developed by Professors Boothroyd & Dewhurst and it is sometimes referred to as Design For Manufacture & Assembly (DFMA). The main principles are:

- Minimise the number of parts by combining or eliminating them where possible.
- Minimise assembly surfaces so that fewer surfaces need processing, and so that all processes on one surface are completed before moving to the next one.
- Design for top-down assembly to take advantage of gravity to assist in assembly.
- Improve assembly access to give unobstructed vision and adequate clearance for standard tooling.
- Maximise part compliance. Design with adequate grooves, guide surfaces and specifications for mating parts in order to reduce misalignment and poor quality.
- Maximise part symmetry for easier orientation and handling.
- Optimise part handling by designing rigid, rather than flexible, parts and by providing adequate surfaces for mechanical gripping.
- Avoid separate fasteners by incorporating fastening into components such as snap-fits. Standardise the fasteners that are used to reduce variability.
- Provide parts with integral self-locking features for easy orientation.
- Drive toward modular design with standard interfaces for easy interchangeability. This allows more options, faster updates of designs and easier testing and service.

Application of these principles will lead to reduced part counts, lower parts cost, shorter design time, higher quality and reliability and reduced assembly time.

6 SUMMARY

Today's increasingly sophisticated markets require product designs that address all aspects of the product's life cycle from raw material to end of life disposal and re-cycling. This makes product design a complex process requiring a wide base of expertise. At the same time the product creation process has to be fully orientated towards commercial success in order to guarantee corporate survival.

The creation of successful life cycle designs requires an approach focussed on costs, cycle times and a variety of Quality Factors. Product Definition is vital as a target setting tool and the designer has to place great emphasis on component selection, variance reduction and design for assembly.

The successful creation of designs in this way, requires a new management process with multi-disciplinary teams containing a wide variety of skills. This is causing quite dramatic changes in management structures with less layers and much less emphasis on functional management. There is a tendency for middle managers to resist the introduction of Concurrent Engineering because of this, but those organisations that can adapt the fastest are the most likely to survive.

FURTHER READING/REFERENCES

[1]   Concurrent Engineering , Carter & Baker, Addison-Wesley.
[2]   Engineering Design for Producibility & Reliability, Priest, Dekker.
[3]   Developing Products in Half The Time, Smith & Reinertsen, Van Nostrand Reinhold.
[4]   Goal Directed Project Management, Andersen, Frude, Haug & Turner, Kogan Page.
[5]   Six Sigma Producibility Analysis & Process Characterisation, Harry & Lawson, Addison-Wesley.
[6]   D.F.A., Boothroyd & Dewhurst, Inc., 138 Main St., Wakefield, RI 02879 U.S.A.
[7]   End-of-life disposal & recycling of electronics, Conference Papers, National Physical Laboratory, Teddington, UK, 16th February 1995.
[8]   The Memory Jogger Plus, Michael Brassard, GOAL/QPC, 13 Branch Street, Methuen, MA 01844, U.S.A.

CHAPTER 4

New Ways of Working at IBM

J. Rook *and* S. Medhat

1 INTRODUCTION

IBM is a leading-edge provider of a diverse range of high technology products, with sites spanning the world. One part of IBM, the Storage Systems Division (SSD), has made significant changes to traditional working methods including:

- Setting up a flat functional organisation - replacing a hierarchy of up to six layers of management;
- Using virtual co-location - where teams in different sites are situated in the same management line. (See Figure 1);
- Implementation of the New Product Introduction (NPI) process to improve business response. (See Section 3).

SSD products are brought to market using the skills of multiple sites as shown in Figure 1. However, this chapter will mainly concentrate on the work undertaken at the IBM Storage Systems Development Group located in Havant UK, which is part of SSD.

A Concurrent Engineering approach has been implemented in the IBM Storage Systems Development Group at Havant, through the adoption of a variety of tools and techniques, such as matrix team working, as shown in Figure 2. They develop storage subsystems and associated technology for IBM's RISC System/6000 workstation products. Products are industry leaders in performance, which is partly achieved by the use of serial link interface technology. This serial interface, now in its second generation as an industry standard, is known as Serial Storage Architecture (SSA) [1].

90

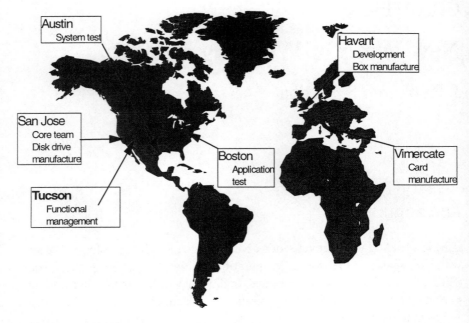

Figure 1   *SSD Multi-site Product Development*

The site at Havant is now owned by a newly formed company Xyratex, after a recent management buy-out. However, approximately 100 development staff still remain as IBM employees who stay as tenants on the site.

2 ORGANISATION

The typical structure for the management of a new product development in SSD, is organised as shown in Figure 2 and will be discussed in more detail in Section 3. The Havant part of the structure is a matrix of product and skill groups with clear boundaries of responsibility. This structure was established to address the increase in the number of projects and the need for reduced cycle time. Figure 3 shows the profile of cycle time and the number of products over the last few years.

Projects at Havant are operated on and delivered using empowered and self-directed team work, with some team members being the champions of their skill area. The ability to work effectively as a member of a team is critical.

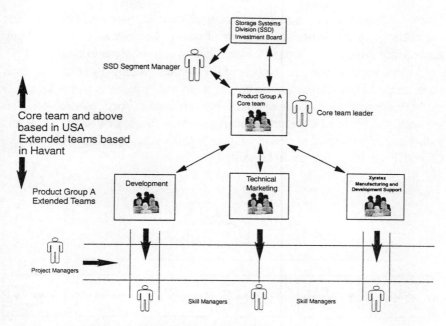

Figure 2   *Typical SSD Project Management Structure*

Cycle time in months

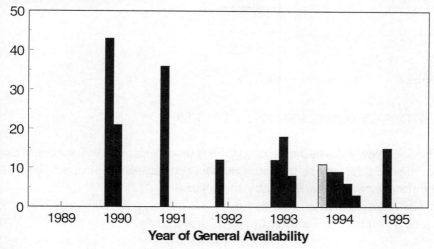

Figure 3   *Past Projects (indicated by each bar)*

92

Using multi-disciplinary teams is not equivalent to forming committees where members often delay decision making. Instead, design teams achieve faster actions through early anticipation, identification and solutions to problems [2].

Skill managers have the responsibility for developing the skills of their department. They also make commitments to the Project managers to supply skills to support the product development schedules. Any resource conflicts are resolved by the Development manager who can prioritise the work and authorise recruitment if necessary. Figure 4 shows the organisational structure of the UK development site in Havant.

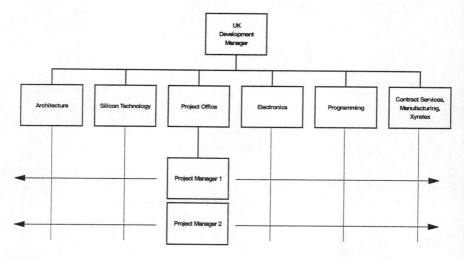

Figure 4   *Organisation at Havant UK*

3 PRODUCT DEVELOPMENT PROCESS

The SSD New Product Introduction (NPI) process aims to provide a consistent development process so that each product program can benchmark itself to comparable programs and industry competitors.

The process includes a number of phases and steps, as shown in Figure 5, which are outlined in Sections 3.1 and 3.2 respectively.

Decision checkpoints are used as mechanisms to review, approve and direct a program at the end of each phase. A phase is considered complete only when the corresponding decision checkpoint for that phase has been signed off. It must be noted here that decision checkpoints are not technical reviews. The checkpoint process is designed so that teams do not engage prematurely in activities for the next phase, until a decision as to whether to further invest resources has been made.

3.1 NPI Phases

3.1.1 Concept phase

The purpose of this phase is to evaluate quickly the attractiveness of a new opportunity and determine if further investigation and therefore investment is justified. Small, cross-functional Proposal Teams are set up chartered by a sponsor (e.g., Segment manager), to explore the viability of potential opportunities. The output of this phase is a 'Product Proposal' and if the Investment Board approves the proposal at the 'Concept Review' decision checkpoint, then the Proposal Team is expanded to form a 6-8 person, dedicated, cross-functional Core Team. These Core Teams will plan and manage new product introductions from conception to end-of-life. They will be physically located in a single work area or in some cases virtually co-located. Teams may participate in a training session to expose them to the concepts of the Core Team process. The training will concentrate on how to apply known resources to building a 'Business Plan', brainstorming on how to address difficult issues, and several exercises to cement team loyalties and facilitate team working styles.

The Core Team directs and integrates the work of functional groups and team members are nominated to represent and lead their functional area. The functional groups provide marketing, engineering, manufacturing and other services. Products that encompass unique issues or uncertainties might include Core Team specialists, such as software, specialised channel marketing, etc. After product launch, the Core Teams remain in place to perform product management activities until the end-of-life of the program.

3.1.2 Planning phase

This phase assesses the potential for success through the creation of a 'Business Plan', 'Product Description', and 'Architecture Specification' by the Core Team.

If the team believes that investment in the opportunity is not justified, then they may deliver a recommendation against investment. The Business Plan will be used by the Board to make an explicit investment decision concerning the opportunity. Detailed development of the product, manufacturing process, and marketing strategy will occur only after an investment decision has been made. The results of this executive decision process aim to provide fewer 'false starts', abandoned programs, or unrealised business cases.

The 'Business Plan Investment Review' is where resources are committed to a product program. A formal 'contract' is drawn up between the Investment Board and the Core Team, which is a statement of the relationship between the team and senior management. It sets out a joint commitment from both parties, with respect to the work plan and its deliverables. Core Teams are responsible for monitoring the terms of this contract throughout the life-cycle of the product and for notifying the Segment manager if a contract infringement is imminent. If this occurs, the Investment Board may choose to review the original justifications and risks of the program, for possible redirection and reinvestment.

3.1.3 Product, process and market development phase

This phase covers all aspects of product design and integration, manufacturing process development, and marketing strategy and planning. Main involvement of the staff at Havant begins here.

The Core Team receives an investment budget which is used to procure services from functional organisations. Functions providing services to Core Teams can staff the work in any manner that satisfies the time, cost and quality requirements set out by the Core Teams. Choices on functional organisation and deployment are solely at the discretion of the functional management team, (although it is expected that the Core Teams will have an advisory role in selection of these Extended Teams). The staff at Havant are one such functional organisation.

Local Project Control Meetings (PCMs) are held in Havant, typically attended by functional management. Development issues are raised and the project's status is evaluated. Such a meeting will highlight issues that are preventing entry to the next development step.

Checkpoints provide a progressive evaluation of the product business outlook as well as a thorough technical assessment as development proceeds.

The Project manager must create and execute implementation plans in order to assure that the completed product meets schedule, function, cost and expense commitments. He negotiates resource commitments with Skill managers to develop and test the product as well as co-ordinating all the business checkpoint tasks assigned by the Core Team leader.

The Havant team includes hardware and software engineers, who work in conjunction with test personnel at the earliest opportunity, in order to allow testing to be efficient and meaningful. This ensures the discovery of problems at the earliest possible stage of development, so that the final testing can be performed with the minimum of defects. Therefore, much effort is put into reviews before hardware and software implementation begins. A 'specification review' which looks at the high level design concepts and a 'design review' which looks at pre-implementation design specifications, are both carried out before any work is implemented. These reviews are placed in the initial project schedule and act as evaluation checkpoints. Participants of the reviews include the following members:

- Author - the specifications writer;
- Moderator - ensures adequate preparation before the meeting, maintains discipline during the meeting and draws satisfactory conclusions at the meeting;
- Reader - any attendee, except the author, to read each section of the material;
- Reviewers (of which at least two are required) - query and probe the subject area.

Representatives with knowledge of the particular aspect of a project under review will attend these working meetings. An 'implementation review' will also occur after a design has been implemented.

Mechanical design is carried out by a sub-contractor. Xyratex, the new owners of the Havant site, are asked to bid for the work.

Simulation is performed at chip, card and system level as appropriate for each design. The micro code may also be simulated before being merged with models of the hardware in the Integration Test step. This step will highlight any major compatibility problems. In general, fault isolation becomes more difficult the higher the level of test that is performed.

Development Teams build prototypes early and as often as is reasonable. This enables them to learn rapidly, minimise mistakes and successfully integrate the work of the many functions involved in the project. Prototypes in this sense mean early mock-ups, computer simulations, test sequences, subsystem models, as well as production prototypes.

Core Teams are not autonomous and the budget provided by the Investment Board is used to contract for resources internally, except where specific exceptions have been agreed during the investment decision making process. In addition, Core Teams are expected to work co-operatively with functions and business unit management, especially in addressing issues of resource constraints or trade-offs.

Two investment reviews occur during this phase. A 'Capital Confirmation Review' is held to assess program status before major capital is committed and the end of the phase is marked by the completion of an 'Announce Readiness Review'. This is used to assess the internal status and competitive position of the program, before committing to announce the general availability of the product.

3.1.4 Validation and ramp-up phase

This phase includes the tasks of design verification, customer evaluations, and manufacturing ramp-up to sustainable production and yield targets. Also, a detailed pilot of manufacturing process implementation activities, on production equipment, is used to confirm manufacturing readiness.

With many products, customer involvement starts early. Prototype units are made available to enable evaluation and feedback on their performance in real application environments.

The Investment Board will review the Core Team business contract at the 'Business Contract Review'. A program post-mortem is carried out and reviewed here, so that best practices as well as lessons learned can be captured and passed on to subsequent projects. The Core Team may be re-configured at this stage (e.g. design engineers are replaced with product engineers) or augmented (e.g. representatives from the Sales and Quality functions join the Core Team).

3.1.5 Life-cycle management phase

This phase considers the management of the program from stable volume production through to end-of-life (EOL). A re-configured Core Team will monitor both market and internal signals (e.g. customer situations and next-generation product transitions), to trigger an EOL implementation plan.

The Core Team is held accountable to the business results which they committed to during the investment decision making process. Therefore they aim to optimise cash flow, which requires a rapid time-to-volume rather than just time-to-market, effective price/volume management during the mature life of the product and efficient management of customer migration and end-of-life inventories.

Prior to the 'EOL Decision Review', the Core Team prepares a detailed EOL plan which will include a customer transition plan, production equipment re-deployment plan, service and support plan and EOL inventory management plan. It is formally presented to the Investment Board, who then prepare to re-deploy Core Teams, other program resources, and production capital.

3.2 NPI Steps

1. *Product Proposal*
   The Proposal Team attempts to quickly determine whether the concept is worthy of further investigation.
2. *Business Plan*
   This is a document prepared by the Core Team in order to allow an investment decision. It is a comprehensive, cross-functional plan expanding on the Product Proposal.
3. *High-Level Architecture & Technology Building Block Assessment*
   The main objective of the step is to define product requirements to a level which instils a reasonably high level of confidence in schedules and resource plans before investing in significant design resources.
4. *Hardware Development & Integration*
   This step translates the Product Description and Architecture Specification into electrical and mechanical designs. A hardware prototype is developed and integrated which is eventually released to the Integration Test and Engineering Verification Test steps, during which it is tested with product micro code.
5. *Test Plan & Cases Development*
   Define the required testing to qualify a product and its components.
6. *Micro code Design & Development*
   Completion of micro code functional definition and detailed coding to support this functionality.

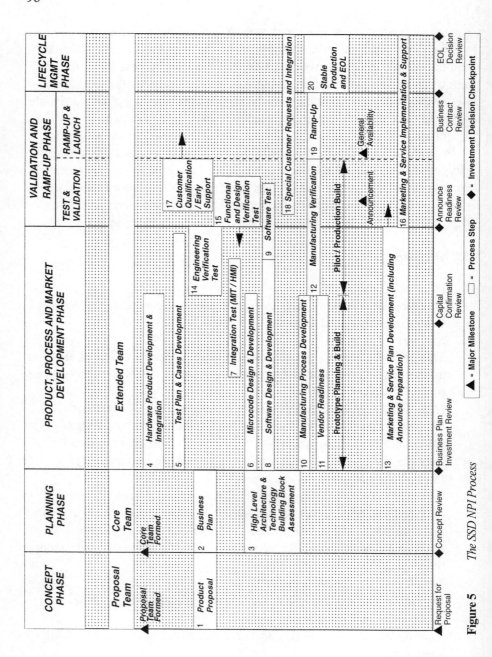

Figure 5   *The SSD NPI Process*

7. *Integration Test*

A Micro code Integration Test simulates the integration, while Hardware / Micro code Integration tests the code with physical hardware.

8. *Software Design & Development*

This represents the interactions between the software development effort and the hardware/micro code development effort.

9. *Software Test*

The hardware product and software product resources are combined in this step. The successful completion of this establishes high confidence in readiness for General Availability, through having completed a full system test, approximating real user environments.

10. *Manufacturing Process Development*

This involves the development of plans and processes for production and production test. At the completion of this step, an end-to-end production process is developed and enabled, ready to begin pilot production runs.

11. *Vendor Readiness*

Vendors are selected and qualified, and parts planning and delivery are completed.

12. *Manufacturing Verification*

Manufacturing demonstrates that the product can be produced on production equipment available, at acceptable quality levels.

13. *Marketing & Service Plan Development*

To support a released product, a detailed 'go-to-market' strategy is developed.

14. *Engineering Verification Test (EVT)*

EVT is a formal prototype evaluation. In addition to classical tests such as EMC, vibration, etc., tests of functionality, compatibility, software, firmware, hardware and error recovery procedures are included.

15. *Functional & Design Verification Test*

This is the evaluation of the product against specification using parts manufactured by the proposed manufacturing processes (in low volume). Fixes to problems found in EVT will be re-tested in this stage.

16. *Marketing/Service Implementation & Support*

The execution and verification of the Marketing and Service Strategy.

17. *Customer Qualification/Early Support*

To validate the product in customer environments and to prepare internal support resources for General Availability (GA).

18. *Special Customer Requests and Integration*

This takes customer requests for special product modifications into consideration and may include process or product changes.

19. *Ramp-up*

Takes production from the General Availability (GA) checkpoint, to maximum achievement of yield, cost, engineering change rate, volume capability and quality targets.

20. *Stable Production and End-of-Life*

Represents long term production and delivery, after the production metrics specified in the Core Team contract have been achieved.

The NPI process as currently adopted in SSD is not designed to be static or permanent. It is destined to evolve and improve over time, as the division continues to practise team-based product development and to draw lessons from both successes and failures.

4 ENGINEERING DATA MANAGEMENT

IBM's commitment to communication is based on its world-wide internal electronic mail system. Virtually all development data and documentation can be transmitted throughout the system. Communication is further enhanced via Local Area Network (LAN) applications that share common data. This enables schedules, engineering and financial data to be accessed by anyone with access to the LAN.

Any problems that might relate to a general problem encountered in, for example, programming methods, can be discussed in on-line forums. This ensures that a problem that might otherwise be ignored, or just discussed between a couple of people, becomes visible. Another person discovering a problem can search for any previous incidence of the problem and gain a much better insight into its cause and resolution.

With many projects running concurrently, there is a need for tight control of schedule as any delay of a critical resource can impact many projects. A PC LAN based planning tool is used at Havant and the Skill managers are responsible for tracking the progress of their deliverables into the main product schedule. The Project manager maintains the overall schedule and holds frequent regular meetings (Project Control Meetings), to review progress against the plan. A 'Project Office' has been formed by the Project managers where tools for managing the information on progress can be developed. They also make a risk assessment against their schedule milestones on a monthly basis.

A recently introduced company-wide software tool, known as Configuration Management Version Control, or CMVC, is used during new product development at IBM. It is designed for use in a networked environment, where software located on a server controls all data throughout the development cycle. Workstations running 'client' software are used to access the information on the server, allowing relevant project data to be worked with by team members. Software and firmware files under development, are maintained in a file system and are managed by a version control system. A relational database on the server maintains all other development data. The organisation of projects within CMVC is hierarchical in nature and an example of this hierarchy is shown in Figure 6.

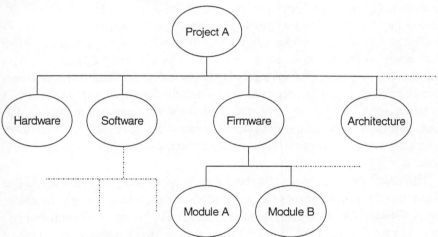

Figure 6   *Configuration Management Version Control Typical Structure*

As can be seen from Figure 6, the top level component defines the project. A hierarchy is then developed to reflect each constituent of the product, in this case, hardware, software, firmware and architecture. These components are expanded as appropriate, for example, modules within firmware. The hierarchy formally defines the areas of product development and each component serves as a storage space to hold specific data. There are no inter-constituent connections.

A problem log is maintained on CMVC and is accessible to any employee with a user ID, to record details of problems found during development. All discourse about the problem is accessible to any user with the required authority. The problems are categorised according to their severity, age, target date, owner, component and originator to name but a few. Severities are classified from one to four, with one being the most severe.

The problem log becomes very powerful as a development tool during test steps. When a developer discovers a problem in a product during a test, it is raised on CMVC. The circumstances under which the problem was discovered and a code dump if required can be recorded. External suppliers also have access to the problem log so that problems that either affect them, or are affected by them, can be dealt with in a similar fashion.

The very significant documentation and clerical task workload associated with change and configuration management becomes a background task automatically managed by the system. The system also provides full traceability by maintaining a complete history of all the changes.

The rapid evolution of the information superhighway has meant that more and more software tools have become available for accessing remote information. A group of software tools known as 'World Wide Web' browsers can be used to access not only textual information but also graphical work. A graphical user interface ensures that these 'web' browsers are simple to use. Data can even be downloaded to the machine that you are using. In addition, web browsers now allow interactive sessions and include animation and sound features.

IBM has exploited the web by providing its own pages of information [3], to allow people from all across the globe to view data relating to new products, press releases, etc. However, these pages can also be set up for internal use only with password protection. This feature has enabled the Havant Development Group to set up its own internal use only web pages. These pages are still under development, but aim to provide access for authorised users to development data from multiple sources. These pages would include:

- Views of project schedules produced on the PC LAN based tool;
- Access to documentation such as specifications, held on a mainframe document library;
- Team constituent details;
- Project details;
- Details of defects on the CMVC problem log.

In essence, the internal web pages bring information together in one location, rather than having to run many pieces of independent software.

5 PROJECT TRACKING AND ASSESSMENT

Methods used previously for monitoring the product development process have, in certain cases included the analysis and prediction of problems found during development. However, these methods have tended to concentrate on one aspect of a product, usually the software element. Much more emphasis is being placed on software within IBM's electronic systems, because software can be more easily and rapidly changed than hardware and can be used to fix either hardware or software problems.

When problems occur during new product development, they are documented and classified as defects using the CMVC software tool. A large proportion of software programming expense can be attributed to the detection and removal of these defects, and the most cost effective removal methods are those that eliminate the defects as early in the development cycle as possible. Various metrics have been proposed relating to software defects, to plan, control and evaluate the software development process, and these enable data to be collected and analysed in a meaningful way [4][5].

A model to predict the number of defects during development will help to provide a more efficient new product introduction in two major ways. In the first instance, the model will give an indication as to whether a project schedule will be met, by predicting where the defect discovery rate will 'tail off'. Secondly, it will give an assessment of product stability at the scheduled completion date, by predicting how many defects have yet to be found if any. In addition, it may provide an assessment of when to economically end testing of a product. This is useful because testing is expensive and needs to be optimised to provide a balance between test coverage and costs associated with the testing. When the rate of defect discovery starts to decrease, and providing that a project's progress is stable (e.g. there are no development problems that are bringing testing to a halt), then the testing can be deemed close to completion. If the project schedule shows more testing resource than is necessary, then there may be scope for reductions in time allocated to testing. The outcome is a more efficient development process and improved product quality. It would be of great benefit if the model developed could account for hardware and software defects.

Reliability growth models have been used in the prediction of software defects [6][7]. Such models require the use of data obtained relatively early in the software development life-cycle, to provide reasonable initial estimates of the quality of an evolving software system.

At IBM, a log is made of all defects found during new product development on CMVC to give accurate and comprehensive statistics for defect analysis and prediction. The most important processes monitored in the development cycle with regard to defect analysis are summarised below:

- Specification Review;
- Design Review;
- Implementation Review;
- Integration Test;
- Engineering Verification Test (EVT);
- Functional and Design Verification Test (FVT/DVT);
- General Availability (GA).

From the defect data available, estimates of future defect numbers for the current project can be made, with initial focus being placed on the firmware, as this yields by far the highest number of defects raised. An IBM internally developed software tool for making these estimates, is being used as an interim measure for the prediction. An example of the curve generated by the tool is shown in Figure 7. It shows the significant differences in time scale between previous and current projects, illustrating the reduced time-to-market of the concurrently developed new product.

The matrix approach to team make-up in conjunction with the recently improved defect logging tool, makes data relatively easy to obtain for use in any predictive modelling. This is due to the projects being divided hierarchically and good communications. Another aspect of Concurrent Engineering that helps in the production of predictions, is the early involvement of downstream activities such as test in the development process. This allows defects to be found at the earliest possible stage and therefore defect data becomes readily available in the early project stages, to provide input to any model for predictions.

6 CONCLUSIONS

To summarise, aspects of Concurrent Engineering that can be seen at the IBM Storage Systems Development Group in Havant, include the following:

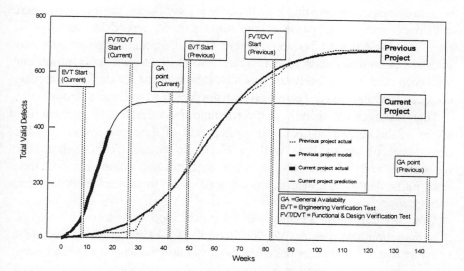

Figure 7   *Firmware Defects from the Start of Integration Test*

- Matrix team structure;
- Team building activities;
- High management visibility;
- The introduction of new team working practices - NPI;
- The co-operation of manufacturing and procurement via physical as well as electronic media. (Quality Engineering techniques such as Quality Function Deployment were used where necessary.);
- CAD use;
- Hardware and software simulation/Virtual prototyping;
- Shared problem log;
- Schedules available to all;
- Early test involvement;
- Use of predictive defect metric tool;
- Communication enhancement workshops;
- Full project reviews and meetings;
- Internal web pages to centralise information.

These processes and activities represent a significant and comprehensive effort to implement Concurrent Engineering. Communication has been regarded as essential, and this is particularly evident during the test steps where developers and testers work on the same part of a product, but may be physically located apart from one another. IBM also benefit from an extremely effective communications system which allows employees to communicate with each other immediately and globally. It is common for staff to have more contact with their Project Team than with the person seated next to them.

High management visibility and communication mean that all staff are fully aware of the current status of the project and of their role within it. The importance of physical as well as electronic communications has been recognised, and subsequently, activities that enhance communication have been implemented to entice functional staff away from their work and into conversation.

One of the most effective development software tools is CMVC (Configuration Management Version Control) [8]. As mentioned in Section 4, software and firmware files under development are maintained in a file system and are managed by a version control system. A relational database on a server maintains all other development data, such as the problem log. Functional as well as management staff are able to understand all outstanding project issues, by having access to a common database where information on problems found during development is stored.

The philosophy of Concurrent Engineering arose out of the need to develop products competitively. One way in which this is achieved is by engineering defects out at the earliest possible stage of development. Enhanced control over the development process will then become apparent and so ambitious schedules are much more likely to be met. Hence, product schedules become a reliable timetable of processes and milestones throughout the project.

In a conventional development process, the discovery of defects at a late stage, possibly because there has been little concern towards designing for the complete life-cycle, dramatically increases the risk of a schedule being broken. The conventional process is very compartmental and does not include aspects such as design for test or design for manufacture at an early enough stage of a project. Finding defects in a late stage of a project may mean that a redesign is necessary. The impact of this will depend on the time that the defect was discovered and the work and resources required to fix it. In general, the later a defect is discovered, the worse the impact. With less control over the development process, the conventional development approach can be classed as unstable. Concurrent Engineering leads to a more stable development process because of the enhanced control [9].

Stuart Marshall [10] the SSD UK Development Manager cited that the use of common tools and predictive defect analysis is essential to manage the schedule exposures. He goes on to say that it is the quality and commitment of the team that will bring a successful product to the market.

REFERENCES

[1]   Information Systems - Serial Storage Architecture SSA-PH (Transport Layer), Document 989D - Revision 3, January 5th 1995, X3T10.1 ANSI Committee, available from : Global Engineering, 15 Inverness Way East, Englewood, Colorado, 80112-5704, US.

[2]   Carter, D., Concurrent Engineering - the product development environment for 1990s, Eddison Wesley, 1992.

[3]   See http://www.ibm.com/

[4]   Chillarege, R. et al, "Orthogonal defect classification - A concept for in-process measurements", IEEE Transactions on Software Engineering, Vol 18, No.11, November 1992, pp943-956.

[5]   Neal, M.,"Managing software quality through defect trend analysis", Managing for quality proceedings of the Project Management Institute annual seminar symposium, Dallas, Texas, 1991, pp119-122.

[6]   Caruso, J. et al, "Integrating prior knowledge with a software reliability growth model", IEEE 13th International Conference on Software Engineering, Austin, May 1991, pp238-245.

[7]   Ohba, M., "Software Reliability Analysis Models", IBM Journal of Research and Development, Vol.28, No.4, July 1984, pp428-443.

[8]   "Configuration Management Version Control" software is produced by IBM and available commercially.

[9]   Proceedings of the International Conference on Concurrent Engineering and Electronic Design Automation, Edited by S.Medhat, Bournemouth, March 1991.

[10]   Marshall, S, "Concurrent Engineering with Multiple Projects and Shared Resources", Proceedings of the International Conference on Concurrent Engineering and Electronic Design Automation, Edited by S.Medhat, Bournemouth, April 1994.

ACKNOWLEDGEMENTS

Credit is extended to the following for his input towards this chapter:
Stuart Marshall - IBM Storage Systems Division, UK Development Manager.

Section 3 of this chapter has been written with reference to the "IBM Storage Systems Division New Product Introduction Handbook", Version 1.0, produced by IBM.

CHAPTER 5

Customer Involvement in Product Specification

A. M. King *and* S. Sivaloganathan

ABSTRACT

The first part introduces and justifies the need for customer involvement in Product Specification. It then prescribes a step-by-step methodology to fully incorporate the customer base into the early phase of a development programme.

The second part demonstrates the methodology in an industrial case-study within the two UK Manufacturing companies. The authors have worked with these companies, and more significantly with their customers, to develop the Product Specification for the next design project. This was accomplished by going out and practising the methodology in a real life situation.

The third and final part gives an assessment of the methodology developed and highlights its strengths together with areas for continued research and application.

1 INTRODUCTION: THE NEED FOR CUSTOMER INVOLVEMENT

During the latter part of the 20th century [1] an increasing number of manufactured goods have changed from being "technology pushed" (where the introduction of new technologies in the designer's control led to the customer specifying what they want as a Solution Concept). This paradigm shift in the development rationale has taken place within almost all domestic/consumer product markets.

Possibly the clearest example of this can be given within the automotive industry. In the pioneering days of motor technology the market was purely "technology driven"; the customers did not know what to ask for, or indeed what to expect, from a new product. It was within this "technology driven" culture that Henry Ford may have uttered the timeless maxim that the customer could have "any colour as long as it was black". The manufacturers held control of new product specification, which only changed as mass manufacturing capabilities became possible.

As the Product Concept of automotive transport became established, so then the balance of control turned towards the customer. Now the customer is able to pull the Solution Concept to fit and fulfil his specific needs and wishes. One example is the development of diesel engines to reduce fuel costs.

This chapter shows a methodology for achieving the aim of "hearing the customer voice" by team participation in the design process.

1.1 A Prescriptive Design Model

Design can be described as a journey from "Problem Land" to "Solution Land". What is needed therefore is a demonstrable methodology of establishing a partnership with customers [2] in a structured way, applicable to all design situations.

It is clear "why" and "what" are needed and so the pressing question now is "how".

Throughout the development of design research, two main schools of thought have developed: the "descriptive design school" and the "prescriptive design school". The former encourages design to be more heuristic with knowledge acquired by the designer used to map out the "best as seen" route to a solution, whilst the latter teaches a formalised prescription of events in the design process.

A prescriptive design model is needed today to develop complex and multi-disciplinary products which are often beyond the ingenuity and scope of an individual engineer.

The customer focused methodology given in this chapter has been developed within the larger framework of a formal, prescriptive Design System called Design Function Deployment (DFD) [3]. The content of this chapter deals with the first of the stages on the prescriptive development path.

These stages can be described as:

- Establishing Requirements & Specifications

- Generating Conceptual Solutions
- Developing Detailed Solutions
- Material & Manufacturing Process Selection
- Production Planning
- Selection of the Optimal Design

The overall model will not provide every detail of every development programme from start to finish, and it is important that the experience and discernment of the team are used to discover which areas of the model to develop and which to leave at a basic level. This paper will demonstrate that a failure to give the first stage proper consideration (establishing the requirements and specifications) will often lead to a poor product performance in the market, or at worst, a design that is not wanted or sold.

1.2 A Guide to Forming a Successful Customer Partnership

Within the activity of Stage One, there are nine prescribed steps as shown in figure 1.

1.3 Develop the Market Insight

Often the commencement of a project can be difficult simply by virtue that little (or nothing) is know about the product. Conversely, if many years of experience have been gathered in one market, there is a danger that a project has no *new* starting point, resulting in the automatic adoption of the previous ideas which may be out-dated.

It is therefore important to write a short "statement of intent" to commence the project. This only need be a few lines as it is often only when a statement is written down that the omissions are seen. This statement will provide the initial direction on the route to establishing the requirements and specifications.

A literature survey can be carried out using libraries to gather background information on the market place, its direction and, more importantly, the customers themselves.

Other places to gather information from include governmental bodies, pressure or campaigning groups and specialist organisations with whom the customers have a direct contact.

Experience has shown that "a fresh pair of eyes" can bring out new aspects of

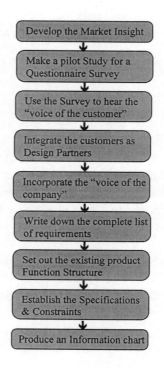

Figure 1   *The Principal Stages in Reaching the Customer*

a problem and so it is *always* advisable to undertake this market insight in some form to gather a panoramic perspective on the market place. Recruiting new members of the development team in this activity can often widen the scope of the exercise because these people will have no constraints to new ideas and will look more widely.

At this early stage, whilst the insight is being established in a particular market, it is always useful to identify and examine other similar products and markets. This exercise may bring in new ideas, diversification of the market, and perhaps, the introduction of a family of products which can be developed for higher market presentation.

In short, adequate insight has to be established before proceeding to the next step, because to treat information as "valuable", one must first be able to comprehend its value.

1.4 Making a Pilot Study for a Questionnaire Survey

A pilot study is a way of testing questions for a major survey and involves writing a small scale questionnaire to be used with approximately 10% of the intended main sample size. The principal purpose of running the pilot survey is to see that the questions and the layout of the information in the questionnaire are clear to the respondents.

The pilot questionnaire should contain *all* the information intended for the main survey and it is important that the respondents are allowed to explain how they understand the questions and are watched as they fill in their answers.

The pilot study is essential to enable ambiguous statements to be modified, omissions filled in, and the confidence for the main survey established. Failure to run a pilot can lead to uncertain replies and, although no survey will be perfect, a pilot study ensures a higher reliability in the answers.

1.5 Using the Survey to Hear the "Voice of the Customer"

The fundamental purpose of the questionnaire is to hear the voice of the customer, to enable their needs to be translated into product functions. It is important that channels of communication are kept open by means of questions to hear their voice, and the questionnaire should focus on the flow of information shown in figure 2.

Special attention should be paid to the layout and administration of the questionnaire and resulting analysis. The following guidelines are based on professional experience for a survey of a sample size ranging from 1000 to 2000 copies.

- Experience has taught that a simple "tick the box" formula will ensure the maximum number of responses. The answer resolution (i.e. number of possible answers) should not be greater than five.

- For clarity of reading use a clear black text font (no smaller than 12 point) on yellow paper.

- Use plenty of space (i.e. leave 30-40% of page blank) to avoid cluttering the page, and give clear guidance through the questions.

- Begin with an explanation of the purpose of the survey, and that the respondent can help improve things.

- Start with questions on why the respondent chose the present product.

- Follow with questions on the present product's performance.

- Suggest new features (gained from the Market Insight exercise) and ask if the respondent will pay for them.

- Ask parallel questions on other equipment/services.

- Background information on the user, both in terms of their personal details and lifestyle, can give useful insight on market trends and help to build the Customer as team Partner. In the questionnaire, these questions ought to be towards the end.

- Cost information is vital to justify major changes in the development programme and reduces the risk in new development. Questions should be succinct such as "Would you pay £....more if....., please answer yes or no".

- Provide room for and encourage additional comments from the respondent.

- Invite them to supply an address if they wish to give further help.

- Write a covering letter to accompany the questionnaire and use a FREEPOST envelope to encourage a reply. A well written questionnaire survey should receive a 10-15% response.

Figure 2   *The Flow of Information Focus in a Good Questionnaire*

1.6 Integrate the Customer as a Design Partner

Traditionally a questionnaire survey is used to establish the broad-band customer requirements, but for detailed knowledge it is important to undertake further investigation.

The involvement of the customer will enable many of the vague or abstract ideas developed earlier in the Design Process to be clarified.

From the questionnaire it will be possible to select a representative sample of the customers who are willing to give extra help. Face-to-face discussions with them will help to break down some of the large terms in the questionnaire in an efficient and effective way.

The customers are the orginator of the requirements of a product and the ultimate judge for its success. It is therefore important to listen to them at all times.

Experience with a questionnaire survey showed that follow-up interviews with a number of respondents enabled a full corroboration of the survey analysis. Where uncertainties were found in replies, the face-to-face interviews helped to clarify the underlying (or effective) issues raised.

More importantly, the questionnaire survey backed up with face-to-face interviews made an excellent mechanism for establishing a team with certain customers.

The next step is to classify the customer requirements in order of priority. To assist this it is best to create an hierarchical structure, namely:

- Primary requirements are the needs which set the strategic direction for the product. Fundamentally they will be broad based and will lack precise information.

- Secondary requirement needs are an elaboration of the primary needs and specify the actions to be taken by the design team.

- Tertiary needs provide the detail necessary for the development of Engineering Solutions.

The number of levels (resolution) in the structure can vary according to he complexity of the product. Benchmarking is a useful technique of comparing a company's own product with the competitors' products. This comparison is made from the customers' perspective. For existing products this process can be carried out in a similar fashion to the rating of customer requirements. A tertiary requirement which is rated as very influential but has not been satisfied by a competitor's product is likely to have a large sales advantage and should receive a high importance rating.

A tertiary requirement having a lower rating (say of 3) when compared to the competitor's rating (say of 5) needs an improvement of (5-3) / x 100% of its present position.

1.7 Incorporate the Voice of the Company

It is important that the product under development should work in harmony with the company's (or companies') corporate policies and in-house practices. The requirements arising out of these considerations should be added to those of the customers' requirements. Specific points for consideration include:

- Bringing the "Voice of the Customer" into harmony with the "Voice of the Company".

- The "Voice of the Company" should sound out the following aspects:

1. Technology Content
2. Capital availability
3. Labour intensive or Automated
4. Manufacture
5. Standardisation

The process of harmonising may lead to bringing the company's voice into harmony with the customer. That is, the practices of the company may warrant changes to accommodate different market and development needs.

It is very important to re-affirm that the customer must form the opinion of the design in the designer's mind. The company's voice must therefore be flexible to harmonise with the customer, rather than vice versa.

1.8 Write Down the Complete List of Requirements

Once the voices of the customer and company have been recorded and understood, the Product Planning can take place. This is where the exact details of what the product should do are stated and prioritised. The priorities are influenced by the following three factors:

- The Customers' Importance Ratings gained from the questionnaire and face-to-face interviews.

- The Sales Advantage, which is given a rating as follows:

    1.0   no advantage
    1.2   moderate advantage
    1.5   significant advantage

- The Necessary Improvement (based on the Benchmarking exercise). The rating is calculated as (target value/current value).

The Relative Importance Rating is obtained by normalising the Absolute Ratings shown above as a scale of 1 to 9.

1.9 Set Out the Existing Product Function Structure

It is important that the requirements are easily translated into meaningful Specifications and Constraints. A simple yet powerful methodology in assisting this is the creation of a Function Structure. This structure is also useful for a later technique called Functional Analysis.

A Function Structure is a breakdown of the intended Product in terms of main, and then more specific, functions. Every product can be structured in this way, by considering the three or four principal functions, and then their subsequent parts. In this way a Function Tree is made. It is often helpful to consider the Primary, Secondary and Tertiary requirements in the earlier phases to provide initial direction on the Function Structure.

1.10 Establish the Specifications and Constraints

The Specifications can now be written in a succinct and precise form, e.g. "the product should operate at speeds of x, y and z m/s" or "the product should not weigh more than x kg". It is obvious that some Specifications will remain qualitative, but the number of such statements should be minimised. The Constraints must all be stated in a clear and measurable fashion. These will often refer to Safety Standards such as "the maximum force exerted must not exceed y N as stated in BS 1234".

1.11 Produce an Information Chart

Once the Specifications, Constraints and Function Structure have been completed, it is possible to lay out the Matrix Information chart. This is an important document which contains all the processed data so far in the development programme, and can be viewed "at a glance" on a single sheet.

2 A DEMONSTRATION OF THE METHODOLOGY

To develop the methodology and prove it by application in an industrial project, a research programme was arranged at the Engineering Design Group, Brunel University with two UK companies [4] within the mobility Healthcare market.

Early "wheeled-chairs" developed from a standard chair design of the time, simply with wheels added. As electrical technology developed, motors were added to the wheels to provide powered motion using battery energy source. In an evolutionary way, the wheeled chair has developed into a sophisticated product with microprocessor control.

Whilst these developments were taking place in the product, changes were also taking place in the market with a considerable increase in the number of active elderly people. This has increased the need for mobility products for the less abled.

2.1 Research Project Objectives

The objectives can be stated as:

- To apply the prescriptive methodology to develop a Customer Design Team to explore the whole personal mobility market.

- To identify by consultation with all possible customer groups (such as elderly, arthritic, stroke etc.) the requirements of these individual groups.

- To research the requirements of all the market segments and assess in terms of product usage, affordability, need and "wanted features".

- To develop the Specifications and Constraints for the next generation of mobility products using a prescriptive design methodology.

- To provide a case-study for critical assessment of the technique used.

2.2 A Customer Partnership

The first thing to clearly establish in the research was the "Problem land" (the situation at present) in the personal mobility market. This clear market definition and customer research was needed to develop a coherent and directed product strategy.

By means of library searches and interviews at a number of locations with Medical Practitioners and Occupational Therapists, the Market Insight was developed.

2.3 Developing the Market Insight

The Market Insight activity produced two main results. Firstly, it provided an excellent basis for the creation of the customer team. This was due in part to the useful contacts and addresses given, but also to the "credibility" gained within the customer base through association with the professional bodies. The value of this aspect of the methodology was greatly strengthened through this advantage.

The second result for the Market Insight was the identification of three distinct market groups instead of the single market specified at the start of the project (figure 3). The first market is the main one associated with wheelchairs and is known as the "disability market". In this market, customers such as those with Cerebral Palsy require an enhanced level of control and seating support where others require only specific needs to be met. However, in addition to that market there is also a range of customers who are considered "less-abled" such as the elderly and those with debilitating (yet otherwise transparent) illnesses such as Multiple Sclerosis. The third (and developing) market is the "enabled" who wish the assistance of a mobility product to support urban transport or recreational activities. The Market Insight exercise ensured that all three

120

markets were known and the first and second developed in the pilot study and questionnaire survey.

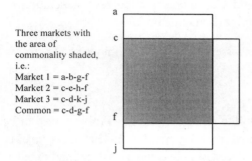

Three markets with the area of commonality shaded, i.e.:
Market 1 = a-b-g-f
Market 2 = c-e-h-f
Market 3 = c-d-k-j
Common = c-d-g-f

Figure 3   *The Common Market*

2.4 The Pilot Study

Once the markets had been established and understood, the research moved to define the market needs with the compilation of pilot questionnaire for the present customers, Occupational Therapists, Dealer Showrooms and the manufacturers and suppliers.

The questionnaires were written with the assistance of professional groups [5] in the healthcare field and, once consultation had been completed, advice was taken on Market Survey techniques from analysts at Anglia Polytechnic University.

The pilot questionnaires were hand-delivered to a selected number of people and discussion held on the questionnaire appearance, question wording and content. The value of this stage cannot be overstated; especially with regard to the customer feedback. Some of the questions were re-phrased to ensure a precise meaning and additional options included to cover all available answers.

The pilot study did not produce the perfect questionnaire, but it did reduce ambiguity and therefore increase the main survey effectiveness.

2.5 The Main Questionnaire Survey

The main questionnaire was printed and sent to 1000 product owners, 30 Occupational Therapists and 20 Dealer showrooms chosen geographically across the United Kingdom.

A 22% response rate was achieved from the users, with a lower rate from the Occupational Therapists and Dealer showrooms of 10%. The survey was analysed and then face-to-face interviews made with a cross-section of the main respondents. The detailed questionnaire analysis is confidential to the research consortia, but a typical graph of results is shown in figure 4.

Complaints with Wheelchair

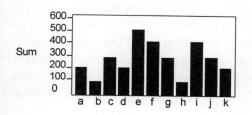

Legend of complaints
a = uncomfortable seating
b = broken frame
c = unreliable
d = punctures
e = difficult to dismantle
f = direction control
g = speed control
h = difficult to recharge
i = expensive maintenance
j = discharged battery
k = rolling back on hill

Figure 4   *Survey Graph of Wheelchair Complaints*

2.6 Face-to-face Interviews

Although the survey gathered a great deal of information, it was not all knowledge. A series of interviews were arranged so that selected respondents could be asked further questions.

Two areas which needed extra clarification were those of safety and reliability. It was clear that these were high on the agenda, but only after asking a number of specific questions: When is it unreliable, where are you when this happens, what have you just done, do you always do "x" before..., etc.?

With the issue of reliability, a long standing problem was resolved by the team interviews. Difficulties with batteries "running out" had often led to the installation of bigger batteries. When this issue of unreliability was investigated further, it was determined that the real source of unreliability was with the charging process, and that was in regard to the user-machine interface, and not the MTBF issue of the components.

The value of a Customer Team was seen again as the survey material could be thoroughly investigated and, therefore, the risk of incorrect data reduced.

After similar interviews with each of the companies, it was possible to piece together a full picture of the current development needs. The use of teamwork enabled the information gathered to become knowledge.

2.7 Product Function Structure

Once the real needs of the users were understood, it was possible to undertake a meaningful benchmarking exercise of the competitors' products, to look for better feature designs to overcome known problems.

Following on from this, a generic function structure was developed to describe each product, and identify the common aspects of each. This was a powerful design tool and enabled the function relationships to be seen in the Function Tree Diagram. Figure 5 shows part of the Function Structure, showing the Power System.

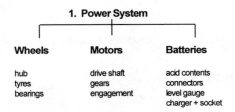

Figure 5 *Part of the Product Function Structure*

The work also identified a number of key parameters in the design which have a high bearing on functions such as safety and stability. To provide design assistance with these, a Mathematical Model was written (in the C programming language) to enable the effects of new designs to be quickly assessed.

2.8 The Flexible Design Strategy

Although largely beyond the scope of this chapter, the work carried out in this project has led to an entirely new development strategy from the one originally intended.

Originally it was envisaged that a new generation of wheelchairs would be developed, but the use of the prescriptive methodology described in this chapter has led to a Flexible Design Strategy.

Such a strategy seeks to maximise Design Re-use by identifying core design components in a number of similar products and developing around these so as to re-use both design concepts and physical components as shown in figure 7.

a)   design for a single market

b)   flexible design
              for a number of markets

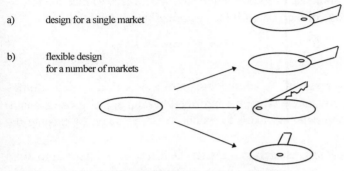

Figure 6   *The Principle of Flexible Design*

3 CONCLUSIONS

This chapter has sought to prescribe and then demonstrate one detailed methodology for realising this need in a practical and workable way. It is often easy to remain at one of two extremes, either theoretic models of a system with no "down to earth" example, or an entirely heuristic teaching which is bound up solely in the experience of the teacher. It has been the aim here to walk the more difficult path between these two, by laying down a workable, and demonstrable set of guidelines.

3.1 Advantages

One clear advantage of the system prescribed has been the ability to see the whole problem (and especially from the Market Insight, the surrounding problem) so as to be able to lay out a longer-term strategy of development. Design is done, ultimately, for one purpose: to make a product at a profit. The use of a prescriptive methodology has enabled this Design Team to identify a long-term strategy where profit and development can be predicted at a lower risk than before.

The findings of the Research Team have been that a prescriptive approach to the process has avoided many common pitfalls and blind alleys, whilst ensuring that the voice of the customer is understood through the triple pass of pilot study, survey questionnaire and face-to-face interviews.

It is often said that "a little knowledge is a dangerous thing" but the three-staged approach of pilot study, questionnaire survey and face to face interviews enabled the knowledge of the "voice of the customer" to be gathered.

3.2 Future Work

The apparent disadvantage of the administration of a prescriptive methodology can be turned to an advantage when the information (and actual process chain) is stored within a computer database to reduce the time taken to execute the methodology.

As part of the on-going research into Design Databases it is intended to build upon this industrial example and develop a computer based model to assist in the process of identifying customer needs. The back-bone to the philosophy will remain the prescriptive path laid down and proven in this chapter; the additional work will enable software technology to provide a faster and more activity-based management system for the whole process chain.

REFERENCES

[1] Clausing Prof. D. "Total Quality Development" ASME Press. 1994. pp 364-71
[2] Cook Dr S. : Technical director of Huntleigh Healthcare. Private Strategy Meeting, 1994.
[3] Design Function Deployment - a design system for the future. Sivalogonathan, Evbuomwan, Jebb & Wynn. Design Studies 16 (1995)
[4] HNE Mobility, Corby, Northants, UK and Penny+Gile Drives Technology, Christchurch, UK
[5] Banstead Mobility Centre, UK; Department of Gerontology, St. James's Hospital, Leeds; Spinal Industries Department, Stoke Mandeville, UK; Department of Occupational Therapy, Roehampton Institute, London, UK.

ACKNOWLEDGEMENTS

The authors wish to thank the EPSRC, UK for the support of this project.

CHAPTER 6

Design Function Deployment - Achieving the Right Cost

S. Sivaloganathan, N. F. O. Evbuomwan, *and* A. Jebb

1 INTRODUCTION

Design for cost is the process of bringing back to the early stages of product development, enough information on costs to enable the designer to use them in decision making. In this way cost is given equal status with other more functional aspects of design and adds to the ability to choose between competing design configurations [1]. Design for cost provides an important basis for minimising manufacturing costs in the most efficient way, i.e. during the design phase. It has been widely cited that about 80% of the cost is committed during the design phase. Therefore it is essential for designers to consider the cost implications of their design decisions when they are made.

Costing is defined as the techniques and processes of ascertaining costs [2]. Expanding on this, costing is the classifying, recording and appropriate allocation of expenditure for the determination of the costs of products or services, the relation of these costs to sales values and the ascertainment of profitability. These costs are determined either (a) historically i.e. after they have incurred or (b) by pre-determined standards. The costing process has been evolving for over 100 years and there is still scope for further development.

In this chapter, cost accountants' methods are reviewed with the intention of identifying their strengths and weaknesses. Design Function Deployment is then discussed with respect to the integration of design for cost. 'Attribute Based Costing' is then advocated as the basis for costing within Design Function Deployment.

2 COST ACCOUNTING METHODS

Accountants analyse the total cost of a product under the following headings:

(i)   Direct material
(ii)   Direct labour
(iii)   Direct expenses
(iv)   Overheads
        (a) Production
                1 Departmental
                2 General
                3 Services
        (b) Administration
        (c) Selling and distribution

Direct material consists of all material that forms and becomes part of the product. It is the material which can be measured and charged directly to the cost of the product. Direct labour is the labour expended in altering the product from raw material to finished product. When analysing direct materials one may often encounter circumstances where the quantity of some of the direct material may be too small for detailed analysis and such direct materials are classified as direct expenses. Direct expenses also include any expenditure other than direct material or direct labour, directly incurred in a cost unit. A typical example of this type of expenses is the hiring of plant or machinery to do a particular job.

Overheads may be defined as the cost of indirect material, indirect labour and such other expenses including services which cannot conveniently be charged direct to specific cost units. There are three main groups of overheads. They are (a) production overheads including services (b) administration overheads and (c) selling and distribution overheads. Production overheads cover all indirect expenditure incurred from the receipt of the order of a product until its completion and subsequent despatch. Typical examples include rate, insurance, indirect labour, power, fuel, consumables, and depreciation of the plant. Indirect material is material that cannot be traced as part of the product.

Indirect labour is defined as labour expended that does not alter the construction, conformation, composition or condition of the product, but which contributes generally to such work and to the contribution of the product. Typical examples include the salaries of supervisors, foremen, inspectors, storekeepers and engineers. Administration overheads consist of all expenses incurred in the direction, control and administration of an undertaking.

There are basically seven application oriented methods in costing. They are :

1. Unit costing
2. Operating costing - applied to services
3. Job costing or Contract costing
4. Batch costing
5. Process costing
6. Operating costing for mass production
7. Multiple or composite costing.

In addition to these, two other kinds of costs can be determined. They are (i) standard costing and (ii) marginal costing.

From the preceding analysis it can be seen that the total cost is derived by adding the direct material, direct labour and direct expense together with the overhead apportioned to the unit. Various terms are used for different combinations of these cost elements. Direct material, labour and expenses are called the prime cost. Prime cost and the production overheads together are called the production cost.

All costing methods revolve around the following steps:

1. Obtain the cost of direct materials
2. Obtain the cost of direct labour
3. Obtain the direct expenses
4. Obtain the production overheads
5. Obtain the administration overheads
6. Add elements 1 to 3 to get the prime cost
7. Apportion the production overheads to each unit produced and add it to the prime cost to obtain the production cost
8. Apportion the general overheads to each unit produced to obtain the total cost

Figure 1 shows the various elements of cost according to the Accounting Method. The fundamental problem with this method is the apportioning of the overheads.This is due to the following reasons:

1. Unlike in the past, where allocating the overheads according to the direct labour hours might have been adequate, automation has reduced direct

128

labour for supervision and setting up in the advanced manufacturing environment. Companies now try to use material and machine hours as the basis for apportioning the overheads.

2. The assumption that allocated costs increase in direct relationship to the volume of product items manufactured is not valid in many cases, since there are many costs that vary with the diversity and complexity of products, and not with the number of products produced.

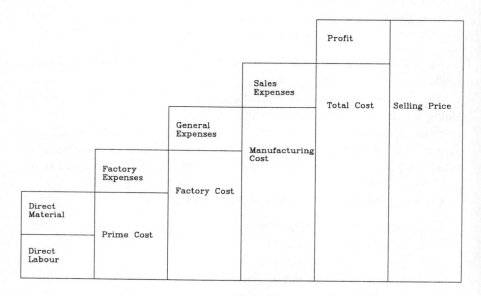

Figure 1   *Elements of Cost*

In addition to these shortcomings in the apportioning of the overheads, the traditional methods suffer from the fact that they are drawn from post event records or from preset standards. In spite of these shortcomings the structure and analysis of the costing system are valuable and formulate the basis for further developments. In particular a system should be developed to estimate the cost elements at the different stages of the design process which provide varying levels of product data. It therefore becomes necessary to define the design model for such a system. Such a comprehensive design system developed in the Engineering Design Centre at City University, London, is briefly described below. Attribute based costing is identified as the basis for costing at the different stages of its design model.

3 DESIGN FUNCTION DEPLOYMENT

Design Function Deployment (DFD) has been developed as a support environment which provides an integrated framework that ensures the concurrent maximisation of design and manufacturing innovation and other life cycle issues during product development. It can be defined as a comprehensive design system, which incorporates the features of a prescriptive design model and associated design methods for the integration of manufacturing, use and other downstream issues into design and thus enabling a "Concurrent Engineering" approach to product/process development.

The design model in DFD provides a systematic approach for the optimal translation of stated (explicit) and latent (implicit) customer requirements and designer intentions into identifiable design functions (specifications and constraints). It also helps to preserve traceability to the original customer requirements throughout the design, manufacture and use stages in the product/process. It ensures that the product is properly conceived at the design stage for manufacture and use [3].

This model proceeds through the following stages :

1. Establishment of customers' requirements and specifications
2. Generation of different conceptual solutions
3. Development of the concepts into detailed solutions
4. Selection of materials and associated manufacturing processes
5. Development of production plans
6. Selection of the optimal solution

The implementation of Design Function Deployment is achieved in three levels where the first level contains the six stages of the design model as described above. The second level contains the tools and techniques which could be used at the different stages of the design model. The third level contains the different databases, rule/knowledgebases, which can be used by the tools in level 2. Design for Cost is such a tool which is located in level 2 with supporting databases in level 3. The structure of Design Function Deployment is given in Figure 2[3].

130

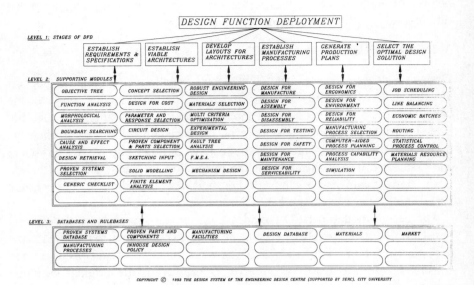

Figure 2   *The Structure of Design Function Deployment*

The product development process within Design Function Deployment is a team activity. It is aimed at adopting a Concurrent Engineering approach to the design process. The design process starts with the establishment of who the customers are, what are their requirements and what are their degrees of importance. The specifications and constraints are drawn up from these requirements in a solution neutral form so that several conceptual solutions can be generated. These solutions can then be costed and evaluated to decide on their further development. In the next stage, the chosen conceptual solutions are further developed into detailed solutions which form the solution space. In the next stage materials and associated manufacturing processes are selected. Production plans are drawn to produce the projected quantities of the system, subsystems and parts. These manufacturing process and production plan parameters are then used to select the optimal solution from the solution space. Cost estimates at each of these stages are necessary to guide the designer in making decisions.

4 SCOPE OF DESIGN FOR COST

The cost of a product can be viewed from two different perspectives. In the first approach, the product design is aimed at minimising the selling price while ignoring the operating cost to the customer in terms of operation expenses and maintenance. In the second approach the total cost of the product, which includes the operation expenses and maintenance, is considered at the design stage and efforts are focussed to minimise this overall cost. Consider the case of photocopiers with replaceable toner powder and toner catridges. The bigger photocopier '*Minolta*' has the facility to replace the toner powder which costs £12 per container. Each replacement will last for about four to six thousand copies on the A4 size paper. A smaller copier '*Canon*' needs the replacement of the complete catridge each time it runs out of toner. The catridge costs about £70 and will last to make three to four thousand copies on the A4 size paper. Of course, there are other factors which influence the selection of the size of photocopiers an institution will be interested in buying. But the operating expenses, which are matters decided during the design stage, can vary considerably as shown above.

4.1 Levels and Attributes of Costing

The details and degree to which a design may be costed is generally dependent upon the complexity of the product and the purpose for which the cost estimate is intended. For example, it would be inappropriate to cost the design of an aircraft in the same way as the design of a pen. At the early stages of design, a pen can be designed and costed to its absolute detail, and thus accurate costing specifications can be drawn which reflect market size, manufacturing, packaging, sales and other issues. Under similar circumstances, it would be impractical and uneconomic to cost the design of an aircraft to such a high level of detail. This is because of the numerous functional features, specialised tooling and materials and many thousands of parts. Although it is possible to cost any part of the aircraft to the detailed level of the pen, the financial returns will be unjustifiable. The level of costing must therefore reflect the scale of the project and the necessary costing details. Cost models normally contain elements of both 'exact' costing and 'estimated' or 'empirical' costing, and their ratio depends on the level of costing required. For example, the costing of a complex object like a ship for tendering purposes would contain a high degree of 'empirical' costing, while a display rack for a garment shop would contain more 'exact' costing.

Different attributes are used at different levels of costing. At the coarse level, the design of a ship may be costed using empirical formulae on facilities provided such as the volume or weight of the vessel, quantity of compressed air or hot water supplied, quantity of floor space provided and so on. It is essential that the major cost contributors are properly identified and methods of estimating the cost, empirical or otherwise, are established for each of these major cost contributors.

Traditionally, all the minor contributors are bundled together as one cost contributor. The empirical formulae used reflect experience, information from the suppliers, and other similar information. In the case of simple objects or when cost estimates with fine details are needed, more information particularly with respect to manufacturing, will become necessary. The preceding analysis suggests that different attributes should form the basis of costing at different levels. In the context of design for cost, these attributes should reflect the different stages in the design process.

4.2 Cost Drivers

The design of a product goes through several stages during the transformation from an abstract brief to a physically realisable or concrete product. At each stage of this development, costs are committed as the design evolves into a physical artefact. Some of the activities and decisions taken at each of the stages generate more costs in the downstream processes; that is, are more sensitive to cost. These are called cost drivers and they occur in all the costing models irrespective of what attribute they are based on. Barnes [4] classifies cost drivers as follows: (a) *volume related* - e.g. direct labour hours, machine hours, direct material costs, and floor space, (b) *transaction related* - e.g. set-ups, receiving orders, material handling, inspections, and scheduling orders, (c) *product related* - physical features: size, weight, surface area, finish, complexity: parts per product, precision, and engineering change orders, (d) *selling, administrative, general* - catalogue pages and changes, utilisation of channel of distribution and capital investment.

This means that a comprehensive design for cost system should allow not only for the various cost models at different levels (and hence the attributes) but also for the identification of cost drivers. Cooper and Kaplan [5] propose that there are a number of cost drivers: for short term variable costs, they advocate that the costs should be traced to the products by volume related cost drivers, while long term variable costs are driven by complexity and diversity. Another important type of cost driver is the number of transactions involved as this often involves inspection or quality checks and some documentation. Other cost drivers being explored include: average number of engineering change orders per month, total number of vendors, number of parts in an average product and the number of customer calls made [6]. These suggest that the design system and the underlying design model should permit and ensure that the product is well conceived, developed, tested and proven before its commital to manufacture.

4.3 Cost Information Databases

It is a fundamental rule in design to use standard components in design to minimise cost. To this extent it is important to have databases of cost information about standard components from various manufacturers. Other important and useful data for cost estimation are method study and time measurement data.

5 ATTRIBUTE BASED COSTING

Attributes are design characteristics which directly contribute to cost. It has been shown earlier that the right attributes should be selected at the different stages of the design model to identify the right cost. French [7] advocates the use of 'Function Costing' at the conceptual design phase to estimate the cost of a new product or system from a specification of its performance. For the effective use of Function Costing, the functions involved must be quantifiable, should be used widely and should have clear cost information. Typical examples are energy provision and storage, normal load carrying capacity and so on. In this context $Cost = CxF$ where $C = Cost\ per\ unit\ function$ and $F = Quantified\ Function$. Often cost is also affected by other parameters relating to other functions and normally to a lesser degree. In these cases $Cost = CxFxW$ where W = weak function of other parameters. Function Based Costing would be very effective when the design elements in the specifications are independent and each requirement is achieved by one or few parameters.

Another attribute that can be considered for costing is that which relates to design features. A feature is an entity with form and function such as a hole, flat face etc.. Design engineers view 'features' in terms of function while manufacturing engineers view features in terms of manufacturing operations and manufacturability. In general, the dos and don'ts that emanate from the manufacturing department contain guidelines on features which highlight manufacturing difficulties and cost. Thus features of a product determined at the design stage, do have cost implications when they have to be manufactured. Hence 'Feature Based Costing' is considered a good candidate for costing a design, once some form (shape, feature, size) is given to it. This will tend to be more relevant as the design process progresses towards the embodiment and detailed design stages.

Activity Based Costing is based on the fact that all activities in a company are to support the production and delivery of the company's products and therefore they should all be considered as product costs. The essence of Activity Based Costing is the understanding of all the activities (that are traditionally called overheads) which consume company resources, and to bring them into the costing system. Such activities include purchasing, establishing vendor relations, receiving, disbursing, setting up a machine, running a machine, reorganising the production flow, redesigning a product and taking a customer order.

In the use of Activity Based Costing, it has been found to support manufacturing excellence by providing information for making strategic decisions on issues like product mix and sourcing; allow designers to understand the impact of different designs on cost and flexibility and then to modify their designs accordingly, as well as supporting the drive for continuous improvement in manufacturing [8]. The early proponents of Activity Based Costing [9] emphasise that it was designed to provide more accurate information about production and support activities and product costs so that management can focus attention on the products and processes with the most leverage for increasing profits.

6 ESTIMATION OF MANUFACTURING COST

The method described here is taken from Malstrom [10]. The method depends on a part explosion list and a diagramatic method is presented to describe the structure of the product to be manufactured. This facilitates the estimating process. This part explosion diagram, sometimes called the Gozinto network, is a means of illustrating each component as part of a manufactured assembly.

These illustrated 'What goes into what' relationships are the basis for the part explosion diagram's alternate name.

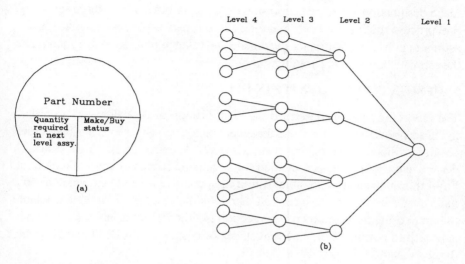

Figure 3   *Part Explosion Diagram*

A part explosion diagram consists of nodes and connecting links. Each node, represented by a circle, is divided into three parts. The upper semi-circle is used to enter the part or assembly number. In the lower left sector the quantity used in the next higher level is entered. The lower right section is used to enter the make or buy decision. Part explosion diagrams are drawn to illustrate the assembly levels. Assembly level 1 refers to the end item or final assembly being manufactured. Level 2 refers to all parts that go into level 1 and in a similar fashion level 3 describes all parts that go into level 2. In drawing a diagram, level 1 is usually placed at the right hand side of the user and all other levels are drawn to the left. Figure 3(a) shows the node details and Figure 3(b) shows the link and node arrangement.

Material costs are estimated based on prices paid for the same materials in the past and on quotations obtained. Quantity requirements drawn from the part explosion diagram are augmented by a fixed amount above the levels to accommodate scrap during manufacture. The first task in estimating the number of labour hours needed to build a product, is to generate a sketch process routing for the assembly under consideration. For machining operations, formulae are used to calculate the running times while for other operations performance standards are used. Process routings should be made for each part that has a make designation in the part explosion diagram. In other words there should be one process routing for each make part in the part explosion diagram. In the estimating phase the estimator must generate a sketch process routing and make the estimate based on this routing.

7 DESIGN FOR COST WITHIN DFD

The preceding sections described the identification of cost elements according to the accountants' methods and then went on to suggest various attribute based costing methods for estimating costs at different levels of detail. The last section described the part explosion method for systematic cost estimation of a product. From this analysis it is possible to find the parts and assemblies which are the main cost drivers. The designer can then look at the different options available to him to bring the cost down. This section is aimed at enunciating a systematic approach to costing at the different stages of design within DFD, based on the discussions in a previous paper [11].

In stage 1 of DFD, the essence here is to examine the possibility of allocating costs to individual customer requirement in a product. This would enable the customers to be more realistic with their requirements as well as appreciate the cost implications of their demands. Designers will also be able to concentrate on individual customer requirement with resources spent proportionately to their importance. Two difficulties, however, arise. The first one is that there are different customers for any particular product and their priorities differ. Secondly, there are different kinds of requirements, namely (i) basic requirements - those that the customer assumes the product will provide, (ii) articulated requirements - those that the customer demands of the product and (iii) excitement requirements - those that if provided would delight the customer. It is hence difficult to have an exhaustive representative list of requirements. However, analysing the customer requirements from a "Design for Cost" viewpoint, will help to improve the definition of customer requirements. Specifications are design elements with 'Target Values' attached to them. They can be used as the input into a "Function Based Costing" system as Advocated by French [7]. The relationship matrix indicating the strong, medium, weak and no relationships should also be used in the Function Based Costing model.

In stage 2 of DFD, different conceptual solutions and their subsystems and characteristics are proposed to meet the specifications. The costing model at this stage should use the subsystem characteristics as the attribute of the cost model. This way the subsystem characteristics and the subsystems which become cost drivers can easily be identified.

In stage 3, the requirements of the subsystem characteristics (which are often alphanumeric or textual expressions) are given the physical structure. Geometric models of the design are built as parts, subassemblies and assemblies. At this stage, the geometric features of the design can be the attributes to base the costing model upon.

In stage 4, materials and associated manufacturing processes are determined. The quantity to be manufactured both for the present and for the future as well as the market size and benefits offered by new materials and technologies are considered. Manufacturing processes and their parameters, together with materials, become the dominant characteristics and hence become candidate attributes for costing models.

In stage 5, the necessary information about the evolving design and manufacture is established. The Activity Based Costing model will be an ideal candidate for the costing model.

The proposed design for cost system within the taxonomy of DFD, is shown in Figure 4.

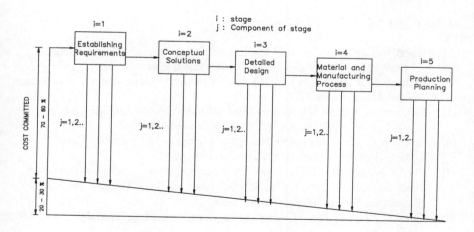

Figure 4   *Cost Drivers in Different Stages of DFD*

138

This figure illustrates that a proper establishment of customer requirements and conceptual design solutions, constitutes about 70% of the cost committed. For each of the costing models proposed for each DFD stage, there will be associated cost drivers, and they need to be identified. The use of the cost models at the later stages of DFD, will also provide feedback information for the earlier stages, for reappraising previous design decisions. The importance of costing models which are representative of the product life cycle, has also been discussed by Tipnis [12]. The paper advocated a framework for a life cycle economic model that can be formulated from early stages of product planning, conceptual design down to manufacturing system design, production planning, service and support.

8 CONCLUSIONS

This chapter commenced with an examination of the traditional accountants' approach to estimating costs within the product development process, as well as showing the limitations of the approach. Design Function Deployment (DFD) was then briefly discussed as a comprehensive approach to product development. Different attributes which can be used for estimating costs at each of the different stages of DFD, were also highlighted. These attributes (function based, features based and activity based), which constitute integral parts of the proposed costing models within the taxonomy of DFD, formed the basis for the enunciation of the "Design for Cost" methodology within DFD.

9 REFERENCES

[1]   Bevan, N et al, "Design to Cost", International Conference on Engineering Design, ICED 89, Proceedings of the Institution of Mechanical Engineers, Vol. 1, pp 415 - 424.

[2]   Owler, L. W. J. and Brown, J. L., "Cost Accounting and Costing Methods", 13th Edition, Macdonalds and Evans, 1974.

[3]   Evbuomwan, N. F. O., "Design Function Deployment - A Concurrent Engineering Design System", PhD Thesis, City University, London, 1994.

[4]   Barnes, F. C., "IEs Can Improve Management Decisions Using Activity-Based Costing", Industrial Engineering, September, 1991, pp 44 - 50.

[5]   Cooper, R and Kaplan, R. S., "How Cost Accounting Systematically Distorts Product Costs", in Burns, W. J and Kaplan, R. S. (eds) Accounting and Management: Field Study Perspectives, Harvard Business School, 1987.

[6]   Drury, C., "Activity-Based Costing", Management Accounting, September, 1989, pp 60 - 66.

[7] French, M. J., "Function Costing: A Potential Aid to Designers", Journal of Engineering Design, Vol. 1, No. 1, pp 47 - 53.

[8] Turney, P. B. B., "Using Activity-Based Costing to Achieve Manufacturing Excellence", Cost Management, Summer, 1989, pp 23 - 31.

[9] Cooper, R and Kaplan, R. S., "Measure Costs Right: Make the Right Decisions", Harvard Business Review, September/October, 1988, pp 96 - 103.

[10] Malstrom, M. E., "Manufacturing Cost Engineering Handbook", Marcel Dekker, Inc., New York, USA, 1984.

[11] Sivaloganathan, S, Jebb, A. & Evbuomwan, N. F. O., "Design for Cost within the Taxonomy of Design Function Deployment", Proceedings of the 2nd International Conference on Concurrent Engineering & Electronic Design Automation, Medhat,S. (ed), April 7 - 8, 1994, Poole, Dorset, pp 14 - 19.

[12] Tipnis, V. A., "Product Life Cycle Economic Models - Towards a Comprehensive Framework for Evaluation of Environmental Impact and Competitive Advantage", Annals of the CIRP, Vol. 40, No. 1, pp 463 - 466.

10 ACKNOWLEDGEMENTS

This work was supported by the Engineering and Physical Sciences Research Council (EPSRC), UK.

CHAPTER 7

Radical Process Improvement through Total Quality Management

L. Jawahar-Nesan *and* A. D. F. Price

ABSTRACT

A plethora of management philosophies dictating different principles, techniques, and procedures for improving quality and productivity have emerged over recent years. This has culminated in cross-functional management concepts such as lean management, concurrent management, agile management, just-in-time, Total Quality Management (TQM) and Reengineering. These all aim to increase productivity and efficiency by holding principles that nurture a culture of customer focus, continuous improvement (radical and incremental), teamwork, empowerment etc., for increased efficiency and productivity. However, the rapid growth of the above concepts and their applications have helped to cause some confusion regarding their linkages and benefits. This chapter examines reengineering as a tool to radically improve business processes within a TQM framework. The combined application of these two concepts, both at project and corporate level, should encourage business organisations, including construction, manufacturing, and service, etc., to be both innovative and productive.

1 INTRODUCTION

Customer demands for improved quality, timely delivery, and reduced costs have resulted in a highly competitive customer focused environment. Quick responses to customer demands, coupled with timely delivery can help companies to achieve competitive advantage. Total Quality Management

(TQM) is a philosophy that uniquely configures people, processes and organisations to respond to the ever-increasing demand of the customer. In order to sustain any improvements and continuously improve processes, TQM requires a number of tools and techniques to be employed when: identifying and solving problems; identifying and analysing processes; and measuring and monitoring effects. Reengineering can be used to analyse work processes and achieve dramatic improvements in business performance. It is a tool to search for new models of organising work. Both TQM and reengineering share a number of common principles and pre-requisites. Using their combined approach, problems prevailing in the construction industry such as unclear project definition, unbuildability, contractual disputes, lack of teamwork, lack of appropriate leadership, and insufficient skilled labour resources can be solved efficiently. This chapter discusses the possibilities of using reengineering as a tool for improving processes within the context of TQM and highlights their application to some of the significant problem areas in construction projects.

2 COMPETITION, CULTURE AND CUSTOMER

Quality and productivity improvement must be the main objectives of any organisation that wishes to become or remain best-in-class. Many top companies have been outperformed in delivering high quality products or services at the right time due to growing global competition [1]. The symptoms of failures are usually: organisational mismatches with the process; short-term goals; reactive management approaches; complacency; and lack of teamwork. These symptoms require management to fundamentally rethink the procedures and principles that they have traditionally followed. This could culminate in a policy of managing for constant change. As the market shifts, economy fluctuates and business environment changes, companies must become flexible enough to cope with these changes; otherwise, they will be outperformed during the course of change. Hence, company culture must nurture the attitude that change is a way to improve and to support a constant improvement process. However, Becky Dunn stated that changing corporate culture is one of the most demanding challenges any company faces, because it takes time and requires far more than "edict from on high" [3]. TQM is seen as one such a philosophy that can be used to solve most of these problems and enables organisations to become best-in-class [4, 5, 6]. It emphasises a culture of continuous improvement and customer satisfaction. Putting this concept into practice will require reengineering existing processes, systems, procedures and the organisation as a whole; changing from function to process oriented, controlled to empowered, protective to productive, hierarchical to lean, and managers to leaders.

3 TOTAL QUALITY MANAGEMENT

Total Quality is widely accepted as "the achievement of customer satisfaction and continuous improvement" [5, 6, 7]. The European Construction Institute's [6] definition states that "Total Quality is a management led process to obtain the involvement of all employees, in the continuous improvement of the performance of all activities, as part of normal business to meet the needs and satisfaction of the customer whether internal or external". In order to achieve the goals of TQM, it has been emphasised that issues such as creating the right environment, effective communications, managing change, empowering people, simplifying processes, measurable aims, and teamwork should be considered by top management. According to Oakland [4], TQM requires total involvement of all employees, total management commitment and customer and supplier working together, combined with objectives, standards and systems which confirm the commitment to total quality. In order to achieve quality, Oakland emphasised the philosophy of "prevention not detection". The conventional approach of "inspection and detection" is replaced by a strategy of "prevention". This concentrates efforts to ensure that inputs are capable of meeting process requirements.

TQM can be helpful in many ways: it collectively integrates people, materials and other resources; it concentrates on improvement towards a common goal; and it achieves efficiency within a business. The British Quality Association [8] stated that "TQM ensures maximum effectiveness and efficiency within a business; secures commercial leadership by putting in place processes and systems that promote excellence; prevents errors; and ensures that every aspect of the business is aligned to customer needs and the advancement of business goals without duplication or wasted effort". In general, TQM as a whole seeks to concentrate on changed culture, management commitment, leadership, customer focus, continuous improvement, employee empowerment, teamwork, training, and communications in order to improve quality and productivity. Table 1 illustrates the main attributes of TQM. Although these attributes seem to be people oriented, the end objective is to improve technology and processes through the improved culture and skill of the people who actually perform the work. Hence, TQM should improve performance, processes, sub-processes, and consequently the end product through the effective use of people and material. Both the British Quality Association [8] and United States Department of Defence [9] stated that one of the objectives of TQM is to "improve processes by the effective application of human resources to the degree to which the needs of the customer are met".

4 REENGINEERING

The term "Reengineering" has gained momentum as a radical improvement management concept since the early nineties. Since its introduction by Hammer and Champy, it has taken different dimensions and definitions under various headings [10]. Some of the terms used in current literature include process improvement, process innovation, business process reengineering and work redesign. Although the terms are different, most of their definitions are very similar in context. Some of the terms and respective definitions have been discussed below.

Reengineering
Reengineering is a radical or breakthrough change in the business process. Reengineering seeks dramatic orders of magnitude as distinguished from incremental improvements in business value [11].

Methods Improvement
Reengineering begins with the customer and works backwards through the process. Everything is geared at what the customer needs or wants and then engineering the process to provide that in the most cost effective way [12].

Business Systems Engineering
A truly "green field" approach that seeks to examine the fundamental processes for a business focusing on the needs of the customer [13].

Process Reengineering
Change in managing processes by re-configuring the value chain from one side

of the business through to the other side [14].

Turner [15] stated that there has been a broad consensus regarding the commonly used term: Business Process Reengineering. Lyons' widely used definition states that reengineering is :

> "the systematic analysis and improvement of business practice
> in a way which furthers the objectives of the business" [16].

Careful analysis of the above definitions suggests that they have been evolved from the original and formal definition by Hammer and Champy [10] which states that:

> "reengineering is the fundamental rethinking and radical
> redesign of business processes to achieve dramatic
> improvements in critical, contemporary measures of
> performance, such as cost, quality, service and speed".

Hammer and Champy insisted on a shift to process-based thinking from task-based thinking. They argued that most problems do not lie in the tasks and the people performing them, but in the structure of the process itself. They emphasised that process-focused management should lead to productivity improvement and cost reduction. Process reengineering, consequently, changes the function of the entire business system in terms of changes in culture, organisation, functional departments, and assembly-line etc. These changes lead to an environment where customer focus is central, managers act like coaches, jobs become multi-dimensional, workers are empowered to make their own decisions and choices, and assembly-line works disappear. Hammer and Champy observed the following changes that result from a company reengineering its processes.

- Work units change from functional departments to process teams.
- Jobs change from simple tasks to multi-dimensional work.
- People's roles change from controlled to empowered.
- Job preparation changes from training to education.
- Focus of performance measures and competition shifts from training to education.
- Advancement criteria change from performance to ability.
- Values change from protective to productive.
- Managers change from supervisors to coaches.

- Organisational structures change from hierarchical to flat.
- Executives change from score keepers to leaders.

The concept of reengineering, as described by Hammer and Champy in their book "Reengineering the corporation", signifies some of the main ingredients required for successful implementation of reengineering. They appear to be the same as those required for TQM: changed culture; management commitment and leadership; customer focus; process focus; employee involvement and empowerment; teamwork; education; and communication. These have also been recognised as the main attributes of reengineering by many experts (see Table 1).

Table 1: Attributes of TQM and Reengineering

Attributes	Total Quality Management						Reengineering				
	British Quality Association (8)	Burati & Mathews (5)	Graves (24)	Allender (21)	Oakland (4)	Pike & Burns (25)	Hammer & Champy (10)	Brittain (3)	Bambarger (26)	Hales & Savoie (17)	Ligus (27)
Changed culture	*	*	*	*	*	*	*	*	*	*	*
Management commitment and leadership	*	*	*	*	*	*	*	*	*	*	*
Customer focus	*	*	*	*	*	*	*	*	*		
Process focus	*		*	*			*	*	*	*	*
Employee involvement and empowerment	*	*	*	*	*	*	*	*		*	*
Teamwork	*	*	*	*	*	*	*	*	*	*	*
Training		*	*		*	*	*			*	
Communications		*			*	*	*		*	*	*

When the reengineering programme is appropriately managed by the effective use of the above attributes, it will improve efficiency, productivity and customer satisfaction. Hales and Savoie [17] stated that the successful reengineering of projects typically lead to breakthroughs in cycle time, quality and cost. Also, the experience of many companies on the application of reengineering proved that they have experienced dramatic improvements in productivity and cost [10, 18, 19].

5 PEOPLE AND PROCESSES: IMPLICATIONS OF TQM AND REENGINEERING

People and processes are the two interdependent factors that should be given due consideration when promoting any business. The people element refers to employees who are involved in or related to a process. A process is a series of activities that produces an output of value to the customer by the use of various kinds of input. A process can be concreting, fabrication, etc. Both product and productivity improvements depend on the structure of the processes that produce them. When processes are structured to align with objectives of the business, then the business would enjoy increased productivity and quality. Deming [20] stated that quality comes from improvement of processes. Hammer and Champy insisted on "process focus" for business improvement. They stated that the problem of productivity lies in the structure of the process itself. However, the "people" who actually perform the process are important for process improvement. It is evident that both TQM and reengineering seek changes in both management and culture. However, according to Brittain [3], Strickland stated that everyone is resistant to change. In this case, the management must consider the "people factor" in order to ensure meaningful change. Otherwise, employees may be startled by the foreboding change. Without employee involvement and commitment, change will not happen. In order to achieve benefits of increased productivity, companies must therefore consider people's ideas, and make them learn how to implement the change process and work in a changed environment.

TQM, inherently, holds the characteristics to improve both the people and process aspects. Simultaneously, it addresses businesses in these two dimensions to improve quality and productivity. However, reengineering seeks

148

first to concentrate on processes and then reorganise the people to accomplish the new work process according to the requirements of the reengineered process itself. In contrast, TQM involves a strong commitment to work with people and incorporates a multitude of techniques to solve problems collectively [21]. On the other hand, reengineering has a strong focus on processes (see Figure 1). According to Brittain [3], Dunn stated that "through emphasis on learning and incorporating the total quality process along with reengineering, people are beginning to shift their view point and adopt new ideas". Figure 1 illustrates that effective implementation of these two concepts together requires concentration on some common factors: changed culture; management commitment and leadership; customer focus; process focus; employee involvement and empowerment; teamwork; training/education; and communications (see previous sections). These factors contribute to activate people to improve their own processes.

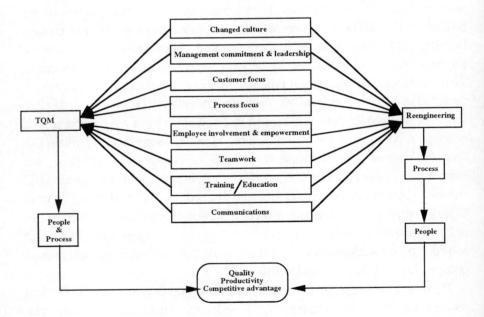

Figure 1   *Relationship between TQM and Reengineering*

Changed Culture

The first requirement for successful implementation is to look at change as a potential opportunity instead of a threat. In order to nurture dramatic change and improvement, it is important to develop a culture that "change is a way to improve and to support a constant improvement process" [22].

Management Commitment and Leadership

Top management must demonstrate real commitment to making employees successful by providing them with the necessary inputs and resources. Commitment also involves follow-up actions to demonstrate a continuing interest. One of the major points identified to help establish a culture conducive to successful implementation is that managers should become leaders. As change occurs, leaders must be able to adjust in flexible manner that is needed to remain successful.

Customer Focus

The requirements by which quality is judged are those of the customer. To be most efficient and effective, the processes and procedures of a company should include the integration of ideas from customers throughout the organisation. This enables the company to satisfy the customer and maintain a competitive edge.

Process Focus

Conventional practices have been in terms of results rather than the process that produces them. However, business improvement as a whole relies on process improvement. It is necessary to understand the key processes that are conducted by the business and both improve and preserve them. By examining the processes, the appropriateness of a process for the business could be judged and redesigned in such a manner as to reach the competitive edge.

Employee Involvement and Empowerment

Every employee has significant potential to improve not only their own functions, but also to co-operate for improvements in other areas. A structure that involves all likely participants in a process enhances the implementation process. In order to create a situation in which the people who actually know what is going on and are in a position to analyse and resolve problems, individuals should be empowered to control their work. Empowering employees is probably the cornerstone of most successful implementation efforts [2].

Teamwork

Individuals working together as a team and towards mutual goals are generally more effective than individuals working alone. Teams established at various levels of an organisation work to improve processes within their control. When properly managed and developed, teams improve the process of problem solving, produce results quickly and economically [4].

Training/Education

In order to maintain a culture of continuous improvement, employees should be educated and trained on skills including leadership, safety, communications, problem analysis and other techniques. Hammer and Champy stated that training increases skills and competence and teaches employees the "how" element of a job. Education increases their insight and understanding, and teaches the "why" element. In a creative and innovative environment, employees at work should possess the skills required to exercise judgement in order to make the right decision. If employees are properly educated, then they could determine for themselves the best way forward.

Communications

In order for innovative ideas to be effectively implemented, there must be an improved communication channel and feedback system to convey ideas for improvement. Communication systems should be designed and managed to benefit those receiving the message, and direct them towards common goals.

6 LINKAGES BETWEEN TQM AND REENGINEERING

Total Quality Management is a mature management philosophy with principles that nurture the improvement of quality and productivity. "Total" itself reflects the holistic characteristics of the philosophy in addressing all aspects of a business, including people, methodology, and technology. It also includes improvement of processes. TQM is neither a tool nor a system to be incorporated in a business, instead it is a culture and philosophy that permeates and navigates an organisation or business towards improved productivity and efficiency. It is a never-ending quest and becomes everyday business. In contrast, reengineering is a concept that focuses upon improving processes. When new technologies and ideas develop, old processes should be replaced by new ones. This contributes to improved productivity and improved cost effectiveness.

Overall, both TQM and reengineering possess similar characteristics and objectives. Both concentrate on improving productivity, quality and performance within a business. Reengineering differs from TQM in its radical reform over processes for dramatic improvement. Hammer and Champy [10] agreed that both TQM and reengineering: share a number of common themes; recognise the importance of processes; and start with the needs of the process customer and work backwards from there. However, there are some fundamental differences that exist between these two concepts. Quality programmes work within the framework of a company's existing processes, whereas reengineering is achieved not by enhancing existing processes, but by replacing them with entirely new ones.

Since the principle of reengineering is "fundamental discontinuous thinking", an organisation can not be reengineered everyday. Brittain [3] stated that reengineering is a one-time effort and has raised the following questions:

- How often a business/organisation should be reengineered?
- If reengineering is a one-time effort, what will be the time interval between consequent reengineering exercises?
- During this interval, what will happen to the reengineered process?
- Will it be continuously improved or left free until the next reengineering exercise?

In order to sustain improvements resulting from a reengineering exercise, management should further look forward to improve the process continuously. Many experts [3, 11, 21] have argued that the approach of "incremental improvement" should follow the process reengineering. According to Brittain, Hal Davis (BST Director/Corporate Quality) stated that "once you finish reengineering a process, the first thing you do is look for ways to improve it".

To further improve the process, he advised the application of quality principles. Also, according to Brittain, Strickland argued that quality efforts can precede the reengineering effort. This lead to the conclusion that a process that is already being run in a quality manner could be reengineered to make a large breakthrough, followed by continuous improvement through quality application principles.

Processes that lead to continual improvements in the timely delivery of products are one of the keys to profitable growth and achievement of competitive edge [22]. Ideally, the continuous improvement approach should comprise both radical and step improvements. If a company progresses with just a strategy of incremental improvement, it may find it difficult to introduce a competitive product on time. On the other hand if, after reengineering a process/organisation or introducing a new product, a company does not look into further improvement of the reengineered process/product, it may not sustain the competitive edge. Both incremental and radical approaches help companies to efficiently undertake the concurrent activities of the new product development process.

The linkages between TQM and reengineering can be defined by the use of three C's: Culture; Continuous Improvement; and Customer Satisfaction. A widely accepted notion is that improved cultural changes (such as changes in people, functional relationships, and method) enhance continuous process improvement. These improvements lead to cost reductions and improved quality, and thus increasing customer satisfaction. TQM seeks cultural changes such as teamwork, employee empowerment, people commitment and involvement etc., whilst continuously improving the business and satisfying the customer. This ensures that TQM addresses all aspects of a business, such as people, technology, and methodology. However, reengineering has been seen as a concept that concentrates on process improvement, which could be considered as a radical improvement activity that delivers a high rate of increased quality and productivity. The following illustrates the direct and indirect linkages between TQM and reengineering.

Direct linkage:
Culture + Continuous Improvement $\longrightarrow$ Customer Satisfaction

Indirect linkage:

Reengineering ⎯⎯⎯⎯→ Radical Improvement

In a broad spectrum, reengineering could well be incorporated as a tool for radical improvement within the process of implementation of TQM. Figure 2 illustrates this concept. Since both TQM and reengineering encourage continuous improvement, they could both be aligned in such a manner that at every interval of the incremental improvement path there must be a radical redesign of process, structure and organisation. This approach strengthens the TQM effort by adding new radical elements for achieving dramatic improvements. The *time* vs *change* and *improvement* curve (Figure 2) indicates that *change* and *improvement* are gradual with time when an organisation seeks incremental improvement in the process of continuous improvement. Quality principles, tools and techniques can be adopted to enhance this process. At some stages, organisations should seek to radically redesign its process, structure, and tasks etc., depending on the trend in the market. The shaded parts of the curve indicate the radical *change* and *improvement*. The curve illustrates that the *change* and *improvement* are slow with time during the incremental stage, where as they are steep with time during radical change. Hence, both the object of TQM and reengineering are achieved by the use of tools of reengineering in the continuous improvement process of TQM.

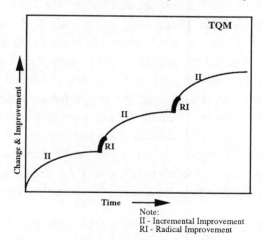

Figure 2   *Reengineering in the Continuous Improvement Process of TQM*

7 TQM AND REENGINEERING THEMES APPLIED TO CONSTRUCTION

There is currently dissatisfaction amongst some UK construction clients regarding the delivery of the completed projects, the quality of service, and predictability of cost. In order to address these problems and improve construction processes, the key drivers to change in the construction industry such as Technology Foresight [28] and Latham [29] have emphasised the need for client focus, teamwork, partnering, contractual arrangements, and empowerment in construction projects. In particular, the Technology Foresight identified four major issues occurring with the highest frequency in UK construction industry: social and environmental factors; cost reduction (including cost in use); international competitiveness; and improved contractual relationships. The following discussion addresses these issues, within the context of TQM and reengineering, with some of the significant problem areas prevailing in the construction industry.

TQM, Reengineering, and People Factors
Construction is an amalgamation of several interdependent sub-processes. Traditional design and alignment of such processes leads to a workforce that is less productive and non-multiskilled. The main attribute of low productivity and poor quality in construction appears to be the failure to realise the importance of the "people factor". It is the individual who exercises a meaningful implementation of the process. In order to create this situation, people should be empowered to control and manage their own work. This should be seen as an enhancement of the processes developed by enhancing a cultural shift which should complement the new processes with new attitudes to work organisation. This is extremely difficult to achieve within construction due to the transient nature of the construction industry. Since both TQM and reengineering focus on improving both people and processes, their combined approach in construction should be efficient if the design and construction processes are developed to focus on customer's expectations and affordability.

TQM, Reengineering and Conceptual Phase of Construction Projects
Reengineering means starting from scratch with a clean sheet of paper and

redrawing the business process from cross-functional teamwork and empowerment perspectives. Reengineering as a 'processes that starts from scratch' has appeared to have a strong correlation with construction projects. Every construction project starts from scratch; it has its own features and processes to be addressed in respect to its type, nature, organisations involved, and client's expectations. The processes and procedures that yielded a successful outcome in one project may not necessarily result in the same outcome in other projects. Blindly following the previous successful procedures may result in the mismatch of product design with that of the process and organisation. Hence, every individual project and its processes should be analysed based on its own constraints and domains. In particular, the conceptual phase of construction projects provides the opportunity to be creative and innovative, where functional objectives, operational concepts, master planning, as well as scheduling and financial goals are analysed and determined based on the project's own requirements. Since the conceptual phase has the most influence on the course of events to follow (i.e. the detailed engineering, procurement, construction, and start-up), the success of these events very much depends upon the decisions made during the conceptual phase. It is in this portion of a project when the seeds of innovation are nurtured [30]. This requires inventive tools and methods to arrive at the right decisions and establish a foundation for a project to be successful. Reengineering could be the best option during the early phase of a project to: design and redesign the product and work processes; design a new project organisation; and design new contractual relationships.

TQM, Reengineering, and Contractual Relationships

Some of the early decisions taken during the conceptual phase relates to contractual strategy, which often influences the working relationships between participants and thus the productivity. The quality of the final product largely depends on the relationship between parties involved in the project. All participants need to share their objectives, resources, and techniques if the process is to be improved. Thus they are required to work jointly towards a common goal, which requires more interactions and interdependence between participants. To avoid misinformation and errors, and at the same time facilitate the exchange of ideas for continuously improving the process, teamwork between participants is essential. Contractual arrangement which places an emphasis on teamwork and partnership is suggested to reduce adversarial situations, improve productivity and reduce cost [28, 29].

TQM, Reengineering, and Fast-track
The competitive business world requires construction companies to do their utmost to beat market competitors. Accelerated project delivery, coupled with cost reduction and acceptable quality are becoming attractive to beat market competitors. Fast-tracking is one of the approaches used by construction companies to reduce their response times in order to cope with the rapid changes occurring in their environments [31]. Fast-tracking requires radical changes in the way construction projects are managed. Moreover, during the simultaneous treatment of both design and construction, many unforeseen problems such as design mismatch with construction and technology, trade interference and work disruptions may arise. Fazio et al [32] revealed that these problems had led to time overrun and productivity loss.

It is believed that genuine consideration of the following, by construction participants, during the simultaneous process of design and construction, can avoid most of the people and process related problems and enable them to deliver projects on time and with an acceptable quality.

- Reducing the cost of the processes involved in design and construction should reduce the cost of the whole project. This depends on the workers performing with a view to improving the process. Empowering the workers to manage and co-ordinate their work activities and co-operating to achieve common goals can help improve the process in terms of time and cost efficiency.

- The use of reengineering concepts at the conceptual phase of a project can enable construction participants to focus on work processes at the early stage of a project. This would enhance constructability of design and produce tasks that are process oriented and customer focused. Once conceptual phase outputs such as project definition, product design, process design, master plan impact, materials supply and utilisation, site utilisation, automation planning etc., are designed through the reengineering tools, the next step is tracking the processes and methods through TQM principles, throughout the project, to achieve further improvements.

- Partnering encompasses many of the characteristics of both TQM and reengineering. It is a key business process for improving teamwork, reducing adversarialism and encouraging innovation through the integrated client-based goals. It will also enhance the implementation process of TQM and reengineering.

- Fast-tracking means rearrangement in design and construction procedures and sequences. It requires radical rearrangement in the way construction industry practices multi-disciplinary design and construction processes. The combined approach of TQM and reengineering in the redesign of processes and organisational structure can improve quality and productivity.

- Goal setting is one of the motivational techniques used in manufacturing and service industries to improve the productivity of the workforce. In construction, the short-term and long-term goal setting techniques could be used to accelerate fast-track processes. Since both TQM and reengineering possess the characteristics of achieving both short-term and long-term change and improvements, their principles can be helpful in achieving goal setting techniques.

8 CONCLUSIONS

Constant change and continuous improvement, both incremental and radical, have to be the driving forces if a business is to be successful. Processes and people are the two important factors, and when properly managed result in changes and improvements that deliver high productivity and better quality. This chapter has argued that reengineering could be effectively used within the framework of TQM for the efficient handling of the process and people towards increased productivity. Incorporating reengineering in the process of TQM nurtures innovation and encourages people to look for ways to continuously improve, and yields both long-term and short-term goals and benefits. In construction, if a project is to be completed using fast-track principles, TQM and reengineering can be used for the efficient design of the facility and work processes. Specifically, reengineering can be adopted at the early stage of a project to achieve highest efficiency in the conceptual phase tasks. This,

158

coupled with continuous-tracking of the processes through the TQM principles in the subsequent phases of the project, should result in improved quality and reduced costs. Construction companies that progress with these concepts should be able to remain innovative and thus be competitive. However, the applications of both TQM and reengineering, requires comprehensive evaluation in order to ascertain the full benefits in construction projects.

REFERENCES

[1]   W. Wrennall, "Productivity: Reengineering for competitiveness," *Industrial engineering*, December, 1994.

[2]   S.R. Sanders, and W.F. Eskridge, "Managing implementation of change, "*Journal of management in engineering*, Vol. 9, October, 1993.

[3]   C. Brittain, "Reengineering complements -Bellmouth's major business strategies," *Industrial engineering*, February, 1994.

[4]   J.S. Oakland, "Total quality management, "*Butterworth-Heinemann*, London, 1994.

[5]   J.L. Burati, Jr., M.F. Mathews, S.N. Kalidindi, "Quality management organisations and techniques, "*Journal of construction engineering and management*, Vol. 118, 1992.

[6]   European Construction Institute, "Total quality in construction," *Stage two report of the TQM Task Force*, Loughborough, April, 1993.

[7]   H. Drummond, "The quality movement," *Kogan Page Ltd*, London, 1992.

[8]   British Quality Association, News letter, *British Quality Association*, London, 1989.

[9]   Department of Defence (USA), "Total quality management," *Master plan*, August, pp. 1, 1988.

[10]   M. Hammer and J. Champy, "Reengineering the corporation-A manifesto for business revolution," *Harpers Business*, New York, 1993.

[11]   R.J. Dixon, et al, "Business process reengineering: Improving in new strategic directions," *California management review*, Vol. 36, Summer, 1994.

[12]   E.E. Torrey, "Observations based on 60 years of success-By the President of H.B Maynard and Co. Inc., *Industrial engineering*, pp. 61, December, 1994.

[13]   F. Turner, "Business systems engineering," *Proceedings of Institution of Mechanical Engineers*, Vol.208, pp. 1-7, 1994.

[14]   M.W. Dale, "The reengineering route to transformation," *Journal of strategic change*," Vol. 3, pp. 8, 1994.

[15]   I. Turner, "Strategy and organisation," *Manager update*, Vol. 6, Winter, 1994.

[16]   L.S. Lyons, "Presentation," *Henley management college*, 1994.

[17]   H.L. Hales and B.J. Savoie, "Building a foundation for successful business process reengineering," *Industrial engineering*, pp. 17-19, September, 1994.

[18]   T.A. Stewart, Reengineering the hot new management tool," *Fortune*, Vol. 128, pp. 41-43, 1993.

[19] D. Crow, "What's all this reengineering stuff about?" *Chemical processes*, Vol. 57, pp. 20-24, 1994.

[20] W.E. Deming, "Quality, productivity and competitive position," *MIT Press*, USA, 1982.

[21] H.D. Allender, "Is reengineering compatible with Total Quality Management?," *Industrial engineering*, September, 1994.

[22] P. Drucker, "Managing the nonprofit organisation," *Harper Collins*, New York, 1990.

[23] R.G. Cooper, "Winning at new products," *Addison-Wesley*, Reading, 1989.

[24] R. Graves, "Total quality-Does it work in engineering management? "*Journal of management in engineering*, Vol. 9, 1993.

[25] J. Pike and R. Barnes, "TQM in action," *Chapman & Hall*, London, 1994.

[26] B. Bambarger, "Carrier transicold teams up with University of Tennessee to implement CLPS," *Industrial engineering*, March, pp. 36-41, 1994.

[27] R.G. Ligus, "Implementing radical change: The right stuff," *Industrial engineering*, May, pp. 28-29, 1994.

[28] Technology Foresight, "Progress through partnership- Report two", *Office of Science and Technology*, HMSO, London, 1995.

[29] M. Latham, "Constructing the team - Joint review of procurement and contractual arrangement in the UK construction industry", HMSO, London, 1994.

[30] A.A. Signore, "Conceptual project planning from an owner's perspective", *Project management journal*, September, pp. 52-58, 1985.

[31] G.A. Britton, "Reengineering the construction process using interactive planning", *First International conference on construction project management*, Singapore, January, pp. 83-90, 1995.

[32] P. Faizo, O. Mosechi, P. Theberge and S. Revay, "Design impact of construction fast track", *Construction management and economics*, Vol. 5, pp. 195-208, 1988.

CHAPTER 8

Virtual Environments for Design and Manufacture

S. Jayaram, S. R. Angster, *and* K. W. Lyons

1 INTRODUCTION

The technology of computer-aided design and computer-aided manufacturing (CAD/CAM) has progressed significantly from the two-dimensional wireframe drafting systems of the 1970s to the parametric and feature-based solid modelers of the 1990s. There are two new technologies which seem to be pushing CAD/CAM into the next generation. These are virtual reality and internet communications. Virtual reality is a relatively new technology which can be regarded as a natural extension to three-dimensional graphics with advanced input and output devices. In very simple terms, virtual reality (VR) can be defined as a synthetic or virtual environment that gives a person the illusion of physical presence. "The exposure most people have to the concept of virtual reality is through reports in the media, through science magazines, and through science fiction. However, to the researchers involved in the actual science of virtual reality, the applications are much more mundane, and the problems are much more real" (1). A good discussion of virtual reality is presented by Machover and Tice (2), and Ellis (3).

The modern product development process calls for rapid design through manufacturing cycles, agile manufacturing systems, adapting designs to suit rapidly changing customer requirements and preferences, use of centralized advanced manufacturing facilities, and outsourcing fabrication. The current suite of product development and rapid prototyping tools are not geared for these new scenarios. Virtual prototyping is a new concept which allows designers to create digital prototypes and evaluate the products thoroughly before a physical prototype is created at a remote location or by a sub-contractor. This significantly reduces time to market and increases the competitiveness of a company. However, the software tools used for virtual prototyping (from traditional CAD systems to virtual reality based manufacturing simulations) are so varied in their architecture that it is almost impossible to create a cohesive set of virtual prototyping tools for use by any organization. This is compounded by the fact that customers and sub-contractors have completely different sets of tools for performing the same tasks.

Many of the issues in virtual prototyping are addressed by virtual reality technology. The ability to view full-scale, three-dimensional models can increase the designer's productivity by allowing a more natural analysis of a model. Virtual fly-throughs of manufacturing simulation can provide a better understanding of the layout of a manufacturing plant. Virtual manufacturing can be used to test and modify numerical control codes off-line and to train workers. Virtual assembly will allow engineers to study ease of assembly and ease of handling. Virtual disassembly can be used to study repair and recycling issues. However, most current systems that have been built to exploit the power of virtual reality are limited in their expandability, customization or usability with current design software systems.

Information that is created and maintained within VR systems must be sharable and capable of being applied and utilized by complementary systems such as computer-aided engineering (CAE) applications. In the case of assembly planning, this tight integration with other design and engineering systems (e.g. CAD and VR functionality with supporting input and display devices, and data exchange) will enable manufacturing engineers to evaluate, determine and select more optimal component sequencing, generate assembly/disassembly process plans, make better decisions on assembly methods (i.e. automated or manual assembly), and visualize the results.

This chapter describes the creation of virtual environments for design and manufacture. The object-oriented architecture of such systems is described using two specific instances as examples. These two systems are VEDAM (Virtual Environment for Design and Manufacturing) and VADE (Virtual Assembly Design Environment). Preliminary results of the use of these systems are also presented. VEDAM is a very general framework for virtual reality applications in design and manufacturing whereas VADE is specifically designed for assembly planning.

2 OTHER RELATED WORK

An overview of the application of virtual reality to CAD/CAM has been presented by Jayaram (1). Several groups have developed systems utilizing virtual reality techniques for early design decisions through the use of virtual fly-throughs, virtual design, virtual assembly and manufacturing simulation. Washington State University has developed a system for the early design evaluation of automobile interiors. This system utilizes Pro/ENGINEER™ models that are brought directly into a virtual design environment. This work was a continuation of a feasibility study that provided successful results in the use of virtual reality for design (4, 5).

Through joint work at the University of Illinois, Chicago, and Purdue University, a prototype, virtual reality based, computer-aided design system has been designed and implemented. The focus of this work is to allow a simplified method of designing complex mechanical parts through the use of virtual reality techniques (6). Work at the Georgia Institute of Technology is focusing on early design changes based on disassembly and servicing criteria (7). The University of Bath in Bath, UK has developed an interactive virtual manufacturing environment. This environment models a shop floor containing a three-axis numerical control milling machine and a five-axis robot for painting. The user can mount a workpiece on the milling machine, choose a tool and perform direct machining operations, (such as axial movements or predefined sequences,) or load a part-program from memory (8). A virtual workshop for mechanical design was developed at Massachusetts Institute of Technology (9). The goal of the project was to develop a simulated workshop for designers to do conceptual design work while having to take into account manufacturing processes. The National Institute of Standards and Technology has developed a Virtual Reality Modeling Language (VRML) interface for a system called VIM, or Visual Interface to Manufacturing (10). This system provides visual access, using VRML, to a database containing manufacturing data, such as three-dimensional models of parts, assemblies and shop floor assembly workstations.

Deneb Robotics (11) has available commercial software for manufacturing simulation, virtual milling, virtual spray painting, virtual arc welding and telerobotics. Most of these systems are precompiled software tools (Deneb). Technomatix Technologies (12) has developed several products in the area of virtual manufacturing. Robcad™ has been developed as a computer-aided production engineering tool for the analysis of robotic applications in a virtual environment. These include welding, laser cutting and painting. Robcad also allows for off-line robot programming and an open system architecture for developing user-specific features and applications. Another product, Part^{TM}, has been developed for computer-aided, numerical control, process planning and programming. Part provides machine tools, machine setup, machining methods and cutting tools. Part will automatically create numerical control programs and process plans given computer-aided design models (Technomatix). Prosolvia Clarus AB (13) manufacturing has created a virtual manufacturing toolkit that is adapted for Technomatix Robcad modules. Users can visualize a manufacturing simulation interactively through the use of virtual walk-throughs of detailed computer-aided design data. This toolkit features an open architecture for users to create their own applications. Clarus is also researching and developing virtual assembly tools (Prosolvia).

Resolution Technologies (14) has created Virtual Mockup^{TM}, a system for fly-through analysis of computer-aided design models. Virtual Mockup provides a set of tools that allows the user to view the models in real-time, query the model's data set, remove objects from the model and save viewing motion paths for playback (Resolution).

Angster (15), Angster and Jayaram (16), and Narayanan (17), created a device-independent, object-oriented, knowledge-based system framework, called the Expert Consultation Environment. This framework gives programmers of computer-aided design applications the ability to add knowledge-based systems to their applications without having to worry about the details of knowledge-based systems programming. Jayaram, et. al., created a prototype of an object-oriented framework to support the integration of multiple software systems used in virtual prototyping of mechanical components (18). The goals of this work included the coupling of software systems which have different data requirements and the demonstration of the system in an industrial setting for the development of electronic spray-cooling components. Jayaram (19, 20) designed an architecture which supports the creation of device-independent, customized product development tools. This architecture included design, user interface, knowledge-based assistant, and virtual manufacturing environments. Woyak, et. al. (21), describe an architecture, called the Dynamic Integration System, which addresses the issues of software integration for engineering applications. This architecture is based on the concept of dynamic variables and dependency hierarchies.

Schroeder et al. (22) of GE Corporate Research & Development, discuss a proposed procedure for designing for maintainability. Along with the accessibility of parts and fasteners, the issues of part path and swept volumes are also addressed. Swept surfaces and volumes are generated by a solid model as it moves through time and space on an arbitrary, time-dependent trajectory. This concept has been applied to the problem of maintainability of jet aircraft engines and "safe" path planning in robot applications (23). According to Kijima and Hirose (24), "the manual handling of objects is one of the basic problems which still exists in the further development (of) virtual reality technology." In this paper, the authors describe the generation of object behavior and application of different models to an object in a VR environment. As stated by Fernando et al. (25), "a common weakness of the existing virtual environments is the lack of efficient geometric constraint management facilities such as run-time constraint detection and the maintenance of constraint consistencies during 3D manipulations."

3 OVERVIEW OF VEDAM

As stated earlier, there are several virtual reality techniques that can assist in the design and manufacturing planning of a product. These include virtual design, virtual assembly, virtual manufacturing, and human-integrated design. A valuable design system for engineers is one that will support all of these techniques, yet be compatible with current parametric CAD/CAM systems. The architecture behind such a system should allow for the expansion and customization of the virtual environments to suit the engineer's needs. An overview of a proposed system, VEDAM, is shown in Figure 1.

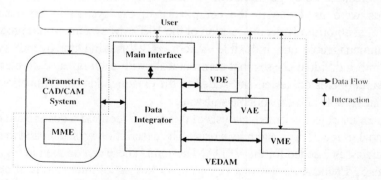

Figure 1   *VEDAM System (26, 27)*

This figure shows the VEDAM system and its components, the Machine Modeling Environment (MME), the Virtual Design Environment (VDE), the Virtual Assembly Environment (VAE), and the Virtual Manufacturing Environment (VME). The MME is an environment that allows the user to create models of the actual machines found in the factory that will be used to produce the product. The VDE is an immersive, virtual reality based design environment that allows the user to view, scale and modify a parametric model designed in a CAD system. The VAE is an immersive, virtual environment that allows the user to analyze the assembly of parts through the direct manipulation of the parts by the user (28, 29, 30). The VME is an immersive virtual environment that allows the user to analyze and develop process plans using a virtual factory that replicates the functions of the real factory.

VEDAM would interface with the parametric CAD/CAM system through the main interface. During a design session, the user would have the option of entering into one of the environments via the main graphical user interface to test designs or manufacturing ideas. All required data from the CAD/CAM system would then be passed into the virtual environments. Upon exiting the virtual environments, the user would have the option of passing data back into the CAD/CAM system. VEDAM, combined with a parametric CAD/CAM system, would provide a complete system for engineers to evaluate potential designs and process plans.

4 OBJECT-ORIENTED ANALYSIS OF VEDAM

Based upon the initial concept discussed above, object-oriented methods were used to create a prototype framework of the VEDAM system. Object-oriented methods were used due to the complexity of the system to be designed. Traditional algorithmic decomposition of the VEDAM system would produce a very unmanageable and inflexible system. By breaking the overall system down into individual classes that communicate with each other, each class can be separately created, tested and added to the existing classes. This allows for very manageable and expandable software.

The initial object-oriented analysis of the system required a description of the functional usage of the system by a user. The description was analyzed to create a set of classes needed for the VEDAM system. These can be seen in Figures 2 through 5. Each figure shows a key group or key set of interactions of classes. Each link between classes represents a dependency between those classes. An open circle represents a "use" relationship, whereas a filled circle represents a "has" relationship.

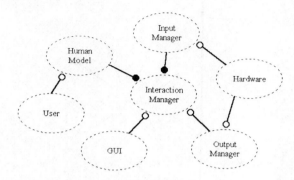

Figure 2   *User Related Classes*

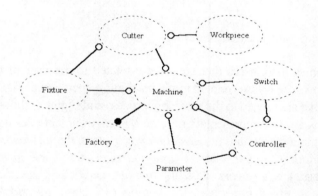

Figure 3   *Machining Related Classes*

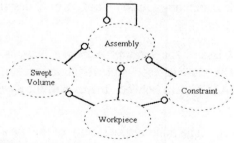

Figure 4   *Assembly Related Classes*

168

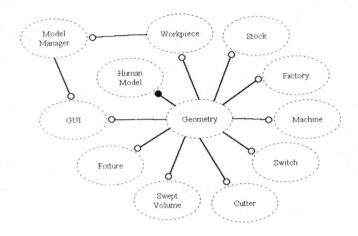

Figure 5   *Geometry Related Classes*

Figure 2 shows the classes that are mostly related to the user of the system. Figure 3 shows the classes related to the machining operations. Figure 4 shows the classes that are related to the assembly process and Figure 5 shows how each class that has a visual representation in the VR environment uses the geometry class. The main classes are the *model manager* class, the *interaction manager* class, the *input manager* class, *output manager* class, the *graphical user interface* class, the *hardware* class, the *human model* class, the *user* class, the *machine* class, the *fixture* class, the *cutting tool* class, the *stock* class, the *workpiece* class, the *assembly* class, the *constraint* class, the *parameter* class, the *switch* class, the *controller* class, the *factory* class, the *process plan* class, the *cutting properties* class and the *geometry* class. A brief description of each class follows.

Interact Manager Class: The *interact manager* class is the center of the VEDAM system. This class is responsible for all interaction that occurs during the use of the VEDAM system between the user of the system and the virtual environment.

Model Manager Class: The main responsibility of the *model manager* class is to act as a data integrator between the virtual reality system and traditional

CAD/CAM systems.

Geometry Class: The *geometry* class is used by many other classes to create a geometric model for visual representation within the virtual environments.

Hardware Class: The *hardware* class is a virtual class for all input and output hardware that will be used by the VEDAM system.

Input Manager Class: The *input manager* class is responsible for getting all the information from the classes derived from the hardware class that represent an input device.

Output Manager Class: Similar to the *input manager* class, the *output manager* class is responsible for passing any information to the classes derived from the *hardware* class that represent an output device.

Graphical User Interface Class: There are two instances of the *graphical user interface* class. One presents the menu system of the main menu to the user at the startup of the VEDAM system. The other responsibility of this class is the presentation of an immersive menu system to the user.

User Class: The *user* class is used to store the information that will be required by the *human model* class.

Human Model Class: The *human model* class is responsible for providing an accurate representation of the user within the virtual environments.

Factory Class: The *factory* class is used to store the information about the virtual factory.

Machine Class: The *machine* class is used to represent a machine in the actual factory.

Switch Class: The *switch* class is responsible for storing information concerning the functionality of a switch, button or lever that is located on a specific machine.

Controller Class: The *controller* class is used by the *machine* class to represent the controller on the machine.

Fixture Class: The *fixture* class represents a fixture used on a machine.

Cutter Class: The *cutter* class is used to represent a cutter that is attached to a machine.

Workpiece Class: The *workpiece* class represents the current state of the part that is being analyzed in the virtual manufacturing environment.

Stock Class: The *stock* class is used to represent the piece of stock that a workpiece is being created from.

Assembly Class: The *assembly* class represents an assembly of parts that was created in the CAD/CAM system.

Swept Volume Class: The *swept volume* class is used to store the volume created by the sweeping motion of parts as they are assembled.

Parameter Class: The *parameter* class is used by many classes for storing parameters and their associating properties. This would include minima, maxima, current values, and methods for changing the values.

Process Plan Class: The *process plan* class is used to store the overall process plan that is created using the virtual manufacturing environment and the virtual assembly environment.

Cutting Properties Class: During the machining process, the various cutting properties are calculated and stored in instances of this class. This class would store such information as cutting time, forces, surface quality and power consumption. All of these properties would be saved as functions of time.

5 INITIAL IMPLEMENTATION AND RESULTS

After completing the object-oriented analysis of the VEDAM system, an object-oriented design produced an initial implementation of the system. The initial implementation provides the user with a virtual manufacturing environment and a virtual design environment. A virtual assembly environment was under parallel development at the time of the VEDAM implementation and current work is being done to integrate the systems. These systems were created on a Silicon Graphics Crimson^{TM} workstation with Reality Engine^{TM} graphics. All classes were developed using C++ and the graphics were created using Performer^{TM} 2.0. The virtual reality hardware used in this implementation include a Virtual Research VR4^{TM} helmet, a Virtual Technologies 22-sensor Cyberglove^{TM}, and an Ascension Flock of Birds^{TM} tracking system with an extended range transmitter and six birds.

The virtual manufacturing environment includes a table-top milling machine, a table-top lathe, and a water jet. These can be seen in Figures 6, 7, and 8, respectively.

Figure 6   *Table-Top Milling Machine*

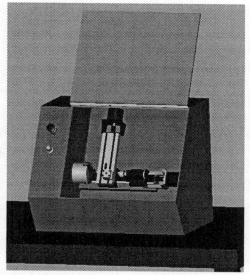

Figure 7   *Table-Top Lathe*

Figure 8   *Water Jet*

All three of these machines are numerically controlled through the use of word address format numerical control codes. A virtual controller on each system provides the functionality of an actual controller. This includes buttons for power, axial movements, setting a floating zero control, loading a part program, and running the program. A graphical user interface (GUI) provides the user with a means for selecting numerical control codes. The virtual design environment is linked directly to the CAD/CAM database through the use of proprietary database interface software. All interaction with the VDE is done through the use of the GUI. The VDE allows the user to select a model to analyze, view parameters of the model, modify the values, and regenerate the model. The user will see the modified model in the virtual environment. A menu of the VDE on the GUI can be seen in Figure 9.

This system uses most of the classes that were identified during the analysis phase. The *model manager* implemented currently supports the transfer of Pro/ENGINEER™ models from Pro/ENGINEER™ into the VEDAM system. These models include the machines, the parts, and the human model. The *interaction manager* has been created to respond to keyboard entry and manages all collision detection between the human model and the environment. The *input manager* controls the input from the keyboard, the data flow of the Cyberglove™ data and the data flow of the Flock of Birds™ data. The *output manager* is responsible for the graphical output of the system to either the monitor or the helmet-mounted display. The *machine* class was created around

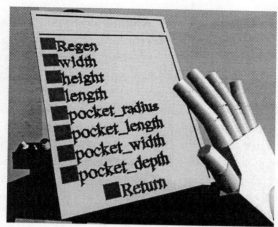

Figure 9   *GUI (Graphical User Interface)*

the development of the table-top milling machine. However, generic methods were written that would provide interactive capabilities for any machine developed in the machine modeling environment. A *graphical user interface* class provides the three-dimensional interactive menu system for the user.

Test cases have been run on all three machines to test the virtual manufacturing environment for accuracy when compared with real machining. The tests involved the comparison of real setup times and machining times with virtual setup and machining times of parts that were created in Pro/ENGINEER™ and Pro/Manufacture™. These times were also then compared to those machining times provided by Pro/Manufacture™ and an estimated setup time provided by a manufacturing expert. The same procedures and numerical control codes were used in both the actual and virtual environments. The tests showed that the virtual manufacturing environment was able to provide a better estimate of setup times for proposed process plans than an expert could. Details of these tests are presented by Angster (26, 31). The virtual design environment demonstrated the ability to view and modify models in a three-dimensional environment through the direct manipulation of a CAD system's database.

6 BACKGROUND OF VIRTUAL ASSEMBLY DESIGN ENVIRONMENT

During a typical design cycle, a product is designed, a prototype is built, changes to the design and production processes are added, and a new prototype is constructed. Once this portion of the design cycle has been completed, the product can then be produced. Often, a part or system is designed without a great deal of consideration for the environment in which it will be assembled. This can be, and often is, a time consuming and expensive method of design. In order to achieve greater efficiency in the design cycle, ways to cut production cost and time to market need to be investigated. With the development of a virtual prototype, both the design and production costs mentioned above can be addressed. If a virtual prototype can be developed for a part or assembly and output to a virtual environment for evaluation, the production costs and time to market could be substantially reduced. The main goal of the VADE system is to provide the designer with the tools necessary to evaluate assembly considerations without the cost and time expenses of the traditional design method. The main problems addressed are as follows:

1. Creation of a virtual environment.
2. Data transfer from a CAD system.
3. Use of CAD assembly information.
4. New information generated.
5. Evaluation and verification of results.

Creation of a Virtual Environment
The creation of the virtual environment involves creating an environment that is flexible from the programmer's standpoint and usable from the user's standpoint. The system needs to be especially robust because the end users of this system will have varying hardware and software setups, system requirements, and computer abilities. Also to be addressed are the ways that the user interacts with the system. Tracking of the head, the hands, and the finger joint angles of the user is needed to provide a believable immersive experience. Once this information is obtained, it is necessary to provide the user with an intuitive way to grab the parts and manipulate them in the virtual environment. Finally, the physical constraints for assembling the component need to be mimicked in the virtual environment.

Data Transfer from a CAD System

The problem of data transfer from a CAD system is of great importance to this research. Once the virtual environment has been created, model data for the user to manipulate, both graphically and computationally, must be obtained. First, the user will need a graphical representation of the parts and assemblies he/she is to assemble and second, the user will also need the relationships between the parts and assemblies so that he/she can accurately assemble the system. The types of information to be obtained from the CAD system are:

1. The graphical representation of the part/sub-assembly.
2. The number of parts/sub-assemblies.
3. The names of parts/sub-assemblies.
4. The final locations and orientations of parts/sub-assemblies.
5. The constraint relationships between the assembled parts/sub-assemblies.

The graphical representation of the part is crucial because the image must convey the representation of the geometry well enough to be believable, but simple enough not to slow down the system. The remainder of the information to be gathered from the CAD system is required to allow the user to assemble the parts in a natural and intuitive way. The main focus of this portion of the research was to find a way to extract this information for the assembly from the CAD system and bring that information into the VADE system. This is a significant problem because the usability of the VADE system relies on how well the user can interact and perform actual assembly operations within the environment. Critical in this interaction is the behavior of the parts in the environment with respect to each other and the user.

Use of CAD Assembly Information

The next step in the creation of the VADE system is the development of methods for using the data obtained from the CAD system. This includes importing a graphical representation of the part, checking final part locations and orientations, checking part constraints, and imposing these constraints on the part's motion. These methods should be efficient and thorough, without sacrificing model or constraint fidelity.

New Information Generated

The purpose of employing a tool similar to the VADE system is to gain some insight into the assembly process in question. Information should be returned from such a system that will allow the designer to improve on the design of the product or the processes involved in assembly. Information generated by the VADE system such as assembly sequencing, swept volume, space requirements, human factors information, etc. should be made available to the designers analyzing the current design and future designers working with subsequent iterations of the assembly. Information of this type should be incorporated into the actual CAD database containing the assembly and the designers' intent.

Evaluation and Verification of Results
Once the user has performed the assembly in the virtual environment, it is important to verify that the results obtained correspond to the results obtained from the actual physical assembly operations. Another consideration in the design of the system is the recording and verification of the part trajectories as they travel along their paths toward final assembly location. Data management and optimization is an important issue since large amounts of data are required to give the designer an accurate description of the path of the part.

7 PROTOTYPE VADE SYSTEMS

In order to gain insight into the functioning of a complete VADE implementation, two prototype systems were created in the process of this research. These initial prototypes were built upon one another to extend the range of knowledge about the desired functionality of this type of application. Problems relating to data extraction in the first prototype were partially addressed by the second. Both initial prototypes used the same system configuration, including the same "core" for creating the virtual environment (28, 30). In the case of these prototypes, the assembly models were generated in Pro/ENGINEER™ and subsequently assembled using the constraint conditions supplied by the Pro/ENGINEER™ interface. No constraint information was implemented except checking the final locations and orientations of the part and gripping, which consisted of attaching the part to the fingertip of a hand model.

The main concern was the development of methods for transferring data between the solid modeling system and the assembler. Several different formats of data transfer were investigated and it was determined that, for this prototype implementation, stereolithography files would be the simplest to generate and convert. Other file formats which were considered include Render™, Inventor™, and IGES. IGES, although the international standard, is too complex for a simple situation where the main aim is to obtain the polygonal representations and relative transformations. Render and Inventor formats contained more information than was necessary for the prototype implementation. Overall, the stereolithography format provided sufficient information and was relatively simple to implement. It was determined through investigating the organization of the stereolithography files that the values required to perform the translation of the part to its final assembled position could be easily extracted from the assembly file, but the final orientation could not.

8 DATA TRANSFER FROM CAD SYSTEM

The interaction between the VADE and CAD systems is of primary importance to the designer. Automation of this link is required so that the user can efficiently and effectively use the system to its full potential. Also, the user must be able to "step out" of the CAD system and "into" the VADE system with minimal effort. The CAD system chosen for this implementation of VADE was Pro/ENGINEER™.

Two methods of interaction between the CAD system and VADE were investigated. The first of these methods of interaction is to start the VADE system from within the CAD system (e.g. - a menu item, etc.). The benefit of this type of interaction is the user's direct ability to access the VADE system with a minimum of knowledge and effort. The second method requires the user to generate variations on the same design or several totally different designs within the CAD system, generate the appropriate data files, and then exit the CAD design environment. The user would then enter the VADE system to evaluate the designs. Ideally, the designers would create an assembly in the CAD system, and with the click of a mouse button, enter the virtual assembly design environment to test the assembly/design assumptions they have made in the creation of the design. Two primary disadvantages of this of interaction are: a) the complexity of the software setup, and b) the prohibitive cost of VR hardware. Hence, for the research presented in this chapter, the interaction with the CAD system was separated from the interaction with VADE.

Data Exchange

The primary information required by the VADE system is a graphical representation of the model itself. The graphical format chosen for the final implementation of the VADE system was the Inventor™ file format developed by Silicon Graphics, Inc. for use with the OpenInventor™ graphics library. It was determined that automated data transfer could be used to obtain the following information.

A. **number** of parts/sub-assemblies - integer value
B. **names** of parts/sub-assemblies - character strings
C. **final locations and orientations** of parts/sub-assemblies - 4 x 4 transformation matrices consisting of 16 floating point values
D. **constraint relationships** between the assembled parts/sub-assemblies.
  1. **type** of constraint - ALIGN or MATE
  2. **geometry information** for mating or aligning object
    a) **axis** - 2 points or 6 floating point values
    b) **plane** - three vectors and an origin point or twelve floating point values
    c) **offset** information about the constraint

The information is obtained through automated transfer from the CAD system using Pro/DEVELOP™, the developer's toolkit for accessing the Pro/ENGINEER™ database. This access is not possible without using this module because of the proprietary nature of the database. Using this toolkit, automated data exchange between the CAD system and VADE proved to be greatly simplified. To perform the data transfer, a new menu item was added to Pro/ENGINEER™'s graphical user interface. This menu item, Pro/VADE, starts the data transfer process using Pro/DEVELOP™'s functionality. For this research, it was decided that only the first level within the assembly tree would be transferred to VADE. Although complex designs are likely to contain several sub-assemblies, any sub-assembly that is part of an assembly can be assembled using the VADE system in a separate session, providing a great simplification to the requirements of the system.

The information first obtained from the CAD system is the number of level-one parts. Next the names of these parts and their model IDs are obtained. Then, the final locations and orientations of each component in the assembly are extracted from the Pro/ENGINEER™ database. The information is stored as a

4 x 3 transformation matrix. The final type of information needed from the Pro/ENGINEER™ database is the constraint information for the assembly operations. An assembly test case was performed to determine exactly what kind of constraint information was needed to perform the assembly in reality and in Pro/ENGINEER™ (29). The information obtained from the CAD system about axial and planar constraints is as follows. For an axis, the two points in space defining the ends of the graphical line representing the axis are obtained. For a planar constraint, three unit vectors and an origin are obtained, thus defining a plane of constraint. The data is then compiled into a VADE data file.

9 PROTOTYPE IMPLEMENTATION AND RESULTS

The first task accomplished in implementing VADE was the object-oriented analysis and design of the system (29). The virtual environment, including stereo viewing and head tracking, was then implemented. The "assembly station" consists of a desk where the assembly operations are performed. This desk was created by taking measurements from a physical desk and generating a crude facsimile using Pro/ENGINEER™. There are also "bins" located in

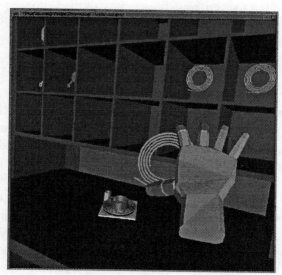

Figure 10   *Visual Environment of the VADE System*

shelving above the desk, so the user can easily select the part for assembly. A ceiling, floor, and walls (one with a window) were added to give the designer a better sense of realism. Once creation of the environment was complete, the graphical representations of the assembly and its parts were imported into the system. Figure 10 shows the assembly environment as viewed by the user.

To allow the user to manipulate the objects within the environment, gripping and releasing of objects were implemented next. To have an intuitive interface between the user and the VADE system, it was desirable to simulate the human hand realistically within the virtual environment. This was accomplished by employing a Virtual Technologies CyberGlove™ to measure the bending and abduction of the fingers. To simulate, or abstract, a "skin" on the fingers, a series of line segments (sensors) were attached to each finger of the hand (29).

This still did not allow the user to perform the assembly operations necessary to complete the assembly, so the constraints and constraining the motion of parts were then created. Along with the constraints on part assembly, a tolerance for final part placement is needed to compensate for the inherent "inaccuracies" of the hardware employed. Unless there is some type of tolerance, exact alignment of axes and planes is impossible. The final step in the development of the system is the recording of the trajectory information of the path the part travels through space to its final location and orientation. This provides the method for verifying the accuracy of the system as well as giving the user the ability to possibly "reserve" space for the assembly process for future iterations of the design.

10 CONCLUSIONS

This chapter has described a framework of a virtual reality based product development system that would aid engineers in the conceptual design and manufacturing process planning stages of a product. Through the use of object-oriented methods, this framework provides an open architecture that allows the easy customization and expansion by the user. The open architecture provides the flexibility to integrate this system with any parametric CAD/CAM system and use any number of input and output devices. The initial implementation demonstrated this feature. As new classes were developed, they were inserted into the system, and only minor changes to the existing classes were made. Also, as more sophisticated methods were developed, only the class that the method belonged to needed to be modified, without affecting the performance of the rest of the classes. These two features allowed a step-wise development of the implementation.

By linking this framework to an existing parametric CAD/CAM system, engineers can immediately obtain the benefits of using virtual prototyping techniques. The analysis of designs in a true, three-dimensional environment, manufacturing the part on replications of the actual factory machines, and the assembly of mating parts are all valuable tools for product development. The initial implementation of this system has demonstrated the feasibility and usefulness of such systems.

The design and implementation of the VADE system were also successful on many points. The prototype VADE system offers designers a unique tool for achieving the goals of a useable, manufacturable, and assemblable product. This system will allow the user to complete the typical design cycle by using the traditional approach of designing an assembly in a CAD system, creating a virtual prototype, making changes to the design based on studies of this prototype, and creating a new and better design based on the information gathered from this cycle.

It is hoped that this work will provide a basis for future work to be performed in the areas of virtual manufacturing and virtual assembly. The successes of the VEDAM and VADE systems provide a starting point for future implementations of similar systems, which will include a more flexible environment, enhanced gripping capability, physics-based modeling, extended constraint functionality, and more complete softzone generation capability.

11 ACKNOWLEDGMENTS

Continued research in this area is being funded by the National Institute of Standards and Technology (N.I.S.T.), Manufacturing Systems Integration Division.

The authors would like to thank Light Machines Corporation for giving permission to model their machines, the proLight Machining Center and the spectraLight Turning Center, for the virtual manufacturing research.

12 DISCLAIMER

Certain commercial equipment, instruments, or materials are identified in this chapter. Such identification does not imply recommendation or endorsement by

182

the National Institute of Standards and Technology, nor does it imply that the products identified are necessarily the best available for the purpose.

13 REFERENCES

[1]   Jayaram, S. (1996a), "Virtual Reality and CAD/CAM," The CAD/CAM Handbook, edited by Machover, C., The McGraw-Hill Company, 1996, p. 410.

[2]   Machover, C. and Tice, S. E. (1994), "Virtual Reality," IEEE Computer Graphics and Applications, Vol. 14, No. 1., 1994.

[3]   Ellis, S. R. (1994), "What are Virtual Environments?" IEEE Computer Graphics and Applications, Vol. 14, No. 1., pp. 17-22, 1994.

[4]   Angster, S. R., Gowda, S., and Jayaram, S. (1994), "Feasibility Study on Virtual Reality for Ergonomic Design," Presented at the IFIP 5.0 Workshop on Virtual Prototyping, Providence, Rhode Island, USA, September 1994.

[5]   Angster, S. R., Gowda, S., and Jayaram, S. (1996b), "Using VR for Design and Manufacturing Applications - A Feasibility Study," 1996 ASME Design Engineering Conferences and International Computers in Engineering Conference, Irvine, August 1996.

[6]   Trika, S. N., Banerjee, P., and Kashyap, R. L. (1997), "Virtual Reality Interfaces for Feature-Based Computer-Aided Design Systems," accepted for publication in Computer-Aided Design, 1997.

[7]   Rosen, D.W., Bras, B., Mistree, F., and Goel, A. (1995), "Virtual Prototyping for Product Demanufacture and Service Using a Virtual Design Studio Approach," ASME International Computers in Engineering Conference, Boston, MA, 1995.

[8]   Bayliss, G. M., Bower, A., Taylor, R. I., and Willis, P.J. (1994), "Virtual Manufacturing," Presented at CSG 94 - Set Theoretic Modelling Techniques and Applications, Winchester, UK, April 13-14, 1994.

[9]   Barrus, J.W. (1993), "The Virtual Workshop: A Simulated Environment for Mechanical Design," Ph. D. Dissertation, Massachusetts Institute of Technology, September 1993.

[10] Ressler, S., Wang, Q., Bodarky, S., Sheppard, C., and Seidman, G., "Using VRML to Access Manufacturing Data," VRML 97: Second Symposium on the Virtual Reality Modeling Language, Monterey, CA, 1997.

[11] Deneb Robotics, http://www.deneb.com/.

[12] Technomatix Technologies, LTD., http://www.technomatix.com/.

[13] Prosolvia Clarus AB, http://www.clarus.se/.

[14] Resolution Technologies, http://www.restec.com/.

[15] Angster, S. R. (1993), "An Object-Oriented, Knowledge-Based Approach to Multi-disciplinary Parametric Design," M. S. Thesis, Virginia Polytechnic Institute and State University, December 1993.

[16] Angster, S. R., and Jayaram, S. (1995), "An Object-Oriented, Knowledge-Based Approach to Multidisciplinary Parametric Design," AIAA 33rd Aerospace Sciences Meeting, Reno, Nevada, January 1995, AIAA 95-0242.

[17] Narayanan, P. (1993), "An Object-Oriented Framework for the Creation of Customized Expert Systems for CAD," M. S. Thesis, Virginia Polytechnic Institute and State University, May 1993.

[18] Jayaram, U., Bakkom, J., Jayaram, S., and Seaney, K. (1996b), "Integration Framework for Virtual Prototyping Software Tools," Final Report, Phase 1, Department of Commerce SBIR Grant, February 1996.

[19] Jayaram, S. (1989), "CADMADE - An Approach Towards a Device-Independent Standard for CAD/CAM Software Development," Ph.D. Dissertation, Virginia Polytechnic Institute and State University, April 1989.

[20] Jayaram, S., and Myklebust, A. (1990), "Towards a Standardized Environment for the Creation of Design and Manufacturing Software," proceedings of the International Conference on Engineering Design, Dubrovnik, Yugoslavia, August 1990.

[21] Woyak, S., Malone, B., Myklebust, A. (1995), "An Architecture for Creating Engineering Applications: The Dynamic Integration System," ASME International Computers in Engineering Conference and the Engineering Database Symposium, Boston, MA, September 1995.

[22] Schroeder, W. A., Langan, J. A., and Linthicum, S. (1994a), "Design for Maintainability," Internal Report, General Electric Corporate Research and Development, Schenectady, NY, URL: "http://www.ge.com/crd/img_and_vis_lab.html".

[23] Schroeder, W. A., Lorensen, W. E. and Linthicum, S. (1994b), "Implicit Modeling of Swept Surfaces and Volumes," Internal Report, General Electric Corporate Research and Development, Schenectady, NY.

[24] Kijima, R. and Hirose, M. (1996), "Representative Spherical Plane Method and Composition of Object Manipulation Methods," Proceedings, IEEE Virtual Reality Annual International Symposium, Raleigh, NC, 1996.

[25] Fernando, T., Fa, M., Dew, P., and Munlin, M., "Constraint-based 3D manipulation techniques within Virtual Environments," Virtual Reality and Applications, edited by R. Eranshaw et. al., Academic Press Ltd., London, 1994.

[26] Angster, S. R. (1996a), "VEDAM: Virtual Environments for Design and Manufacturing," Ph. D. Dissertation, Washington State University, December 1996.

[27] Angster, S. R., and Jayaram, S. (1996c), "VEDAM: Virtual Environments for Design and Manufacturing," Symposium on Virtual Reality in Manufacturing Research and Education, Chicago, October 1996.

[28] Connacher, H. I., Jayaram, S., and Lyons, K. (1995), "Virtual Assembly Design Environment," 15th ASME International Computers in Engineering Conference, Boston, MA, September 1995.

[29] Connacher, H. I., "A Virtual Assembly Design Environment," M. S. Thesis, Washington State University, Pullman, WA, May 1996.

[30] Connacher, H. I., Jayaram, S., and Lyons, K. (1997), "Virtual Assembly Using Virtual Reality Techniques," accepted for publication in Computer-Aided Design, 1997.

[31] Angster, S. R., Jayaram, S., and Hutton, D., "Test Cases on the Use of Virtual Reality for an Integrated Design and Manufacturing System," 1997 ASME Design Engineering

184

Conference and Computers in Engineering Conference, Sacramento, CA, September 1997.

[32] Bennet, G. R. (1995), "Virtual Reality Simulation Bridges the Gap Between Manufacturing and Design," Mechanical Incorporated Engineer, April/May 1995.

[33] Jaques, M., Strickland, P., Oliver, T. J. (1995), "Design by Manufacturing Simulation: Concurrent Engineering Meets Virtual Reality," Mechatronics, 1995.

This chapter includes work by United States Government employees, which is therefore in the public domain.

PART III

MANAGEMENT OF CONCURRENT ENGINEERING

CHAPTER 9

Making Teams Work

J. R. A. Greaves

Teamwork has become the management panacea for organisation ailments. The language of the sports coach has infiltrated the business environment to make us all develop teams - cross functional teams, virtual teams, project teams. This chapter exposes some simplistic myths and provides hard earned practical lessons in how best to tap some of the 70% of people's capability typically unused by our major companies, through working together. It focuses on a particular project-based UK defence/electronics company, who were faced with a downward trend in financial performance, driven by a cost plus, consequence-free culture and demotivated staff. The company had the potential to be world class, with technological capabilities second to none, and leadership in core markets. They needed a radical rethink of their ways of working, based on teams.

1 WHAT IS TEAMWORK?

Teamwork may be accepted wisdom but there are many levels of it, and many organisations settle for a nominal level. They just co-locate individuals from skill areas into project groupings and think the problem is solved.

Figure 1, overleaf, shows part of a Maturity Profile describing levels of teamworking from Innocence (how do you spell it?) to Excellence (is there any other way?). Each indicates varying degrees of sophistication in working together:

Innocent: -   each skill area contributes to project tasks, but asks no
                       questions, with the result that there is no synergy;

Making Teams Work

Maturity levels of teamworking

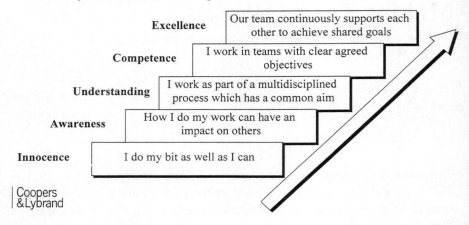

Figure 1   *Maturity Levels of Teamworking*

Awareness: -   realisation of the internal customer/supplier network;

Understand: -   some process literacy with the notion that 'we are all part of something which is bigger than my area' such as a development process, contract delivery process, or bid process;

Competent: -   forming teams with agreed objectives made explicit at launch;

Excellent: -   objectives shared, all activity driven to meet them
    -    knowing when it is right to engage in a creative joint process (e.g. planning, problem solving)
    -    harnessing all capabilities
    -    mutual support, challenging positively
    -    people skills becoming pre-eminent

2 WHY DOESN'T TEAMWORK HAPPEN?

A causal analysis of the problems faced by this company pointed to some familiar issues including functional silos, conflicting performance measures, lack of shared needs/objectives, and lack of trust and responsibility, with problems shifted instead of solved. These all relate to poor teamwork, and the lack of awareness that to the outside customer it doesn't matter which part of the company has let him down, it's the whole company.

How did this situation arise? Too often our companies are set up specifically to prevent good teamworking. Warner Burke and George Litwin at MIT (see reference 1) developed a system model of organisations, which identifies all the basic elements and the interactions between them. It indicates that there are three strata in the organisation model (Figure 2). In this case all three contained constraints to team behaviour.

Making Teams Work

Constraints on teamworking

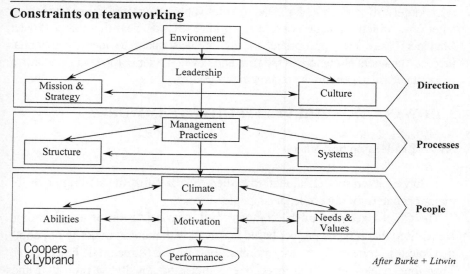

Figure 2   *Constraints on Teamworking*

(a)   **Direction**
      The leadership revealed a residue of autocratic role models with a consistent command and control management approach.

(b)   **Process**
      A structure had been created with market-facing divisions trading formally with engineering and manufacturing functions.   The Chief Executive was the only legitimate point to solve conflicts across projects.

(c)   **People**

The general motivation was to keep noses clean and avoid failure. Too many people were primarily interested in time filling or engaging in technical 'hobbies'.

All of this was wrong for teamworking *but changing any one is not enough, since other aspects conspire against it to stabilise the situation back to the status quo.*

A detailed analysis of people's abilities was revealing, particularly when looking at the population of Project Managers. It showed that only 10% had an adequate set of skills, knowledge, and attributes. Some competencies were good enough, such as negotiation for all the internal trading activities, but others were largely absent - coaching, leading, motivating, communication. It is these latter ones which facilitate good teamworking. It was clear that many Project Managers were not appropriate and action was taken to identify potential Project Managers elsewhere in the firm and develop their competencies within a reorientated management development programme.

3 HOW DO YOU ACHIEVE EFFECTIVE TEAMWORK?

3.1   Leadership and Vision

The Burke Litwin model, alluded to earlier, indicates that the starting point for major change is the area of leadership and vision.

The Chief Executive of the company had a strong vision of teamworking. He realised that it was a project-based business, with the vast majority of people dedicated to one project for months, but it was structured functionally. Therefore, a new ethic was needed encompassing loyalty to project teams, active problem solving, a sense of responsibility for the project output, a culture of admitting failings and honesty, and the engagement of crisis-style energy before crises occur. He knew it needed a new style of team leader with different skills.

A concept of 'project focus' was developed and communicated aggressively through presentations, team briefings, exhibitions and newsletters. The message to all staff was to identify with their project ('wear the team shirt'), to adopt an agree/commit/achieve mentality, and to think along process lines.

Of course, all this had to be backed up by management protocols, roles, responsibilities and supporting systems to make teamwork more than just a fanciful vision.

3.2 Organisation and Structure

Teamwork requires an appropriate organisation environment to flourish. There is no single ideal set up, since much depends on the need for multiplexing, the duration of projects, skill width, and the complexity and resource profile of the work. Of the major organisation types found in project-based businesses, the basic functional (task and resource authority with line managers) and lightweight matrix (some tasks in teams with little authority) structures are typically least conducive to project teamworking. A project-based matrix, where task management is carried out through the project axis, and resource management is done through both functional and project axes depending on resource type, proved optimum in this case. This gave the advantages of good cross fertilisation of skills, clear responsibilities, and resources provided to projects for agreed durations.

Projects themselves need a basic structure too, since teamworking doesn't just happen as soon as goals are shared.

(a) **Co-location:**
The answer here was to identify project war rooms with some staff co-located such as the key system engineer and project leader. Bases in skill areas were retained because of the need for access to IT and knowledge pools, especially in engineering and where capital assets were required (e.g. test).

(b) **Teams within teams:**
A network of teams starting with a core team was instigated for major projects (Figure 3). The core team is ideally formed at the bid phase and is always cross functional. It holds responsibility to plan and accept commitments on behalf of the functional areas. Although not all members are full time, each one owns some of the project ('milestone owner'). Each of these milestones has its own team underneath.

(c) **Ways of working:**
New processes for planning top down, reporting upwards, and project control were created. The processes for problem solving, decision making and planning took place within the team at each level.

Making Teams Work

Project structure

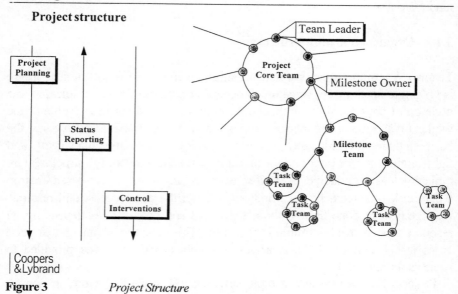

Coopers
&Lybrand

Figure 3 *Project Structure*

Working in this new way meant that 40% to 60% of staff doing work for a project were on this team, compared to between 4% and 10% previously.

3.3 Roles

The key shift here is from Project Manager to Team Leader, whose prime roles are to motivate and coach. There is a surprisingly low requirement for detailed technical knowledge, requiring greater trust in other colleagues' solutions; a trust soon rewarded by commitment and realism in proposed responses to contract needs.

3.4 Processes

In most engineering companies such as the one under consideration here, process thinking is a foreign concept, indeed a process is seen as merely something physical like the assembly of printed circuit boards. The results are management processes conspicuous by their absence, projects that are never really launched, with slow ramp up and unclear objectives and deliverables. Deadlines are only achieved by last minute firefighting.

There are, however, new processes required for teams to work. One of them is that for launching a project:

Step 1:   Secure the sponsorship; get board level commitment to the project, and have a specific champion or owner of it.

Step 2:   Appoint a leader - emphasising this key role.

Step 3:   Select and mobilise a core team - incorporating teambuilding, agreeing objectives, constraints, planning, resource needs, and modus operandi.

Step 4:   Link into the company management processes - progress reporting, cost control, resource planning, manufacturing scheduling, and procurement cycles and processes.

All too often these steps are assumed to have taken place, without formal recognition. Some company management processes have to be completely reengineered to enable good teamworking, such as that for Task Loading:

Old	New
Projects are broken down by a central planning department into work packages, using a sophisticated IT planning tool;	Core project team prepares a milestone plan for the project with identified skill needs;
Each work package is subcontracted to departments by an internal order system, thus splitting it all into functional skill boxes;	Agreed resources are hired onto the project from skill areas; Teams are brought together rather than work parcelled out;
Responsibility for activities firmly within the department structure;	Internal agreements are struck between people to deploy resources for an agreed time/output.
The project manager has no control or visibility.	

194

This results in less multiplexing of tasks, which more than compensates for minor inefficiencies in resource utilisation. It also breaks the functional stranglehold and manipulation to keep people busy contrary to project needs.

3.5   Team Based Planning

The actual planning of project work can negate the best efforts to promote teamworking, by removing all potential for shared commitment to agreed goals. The case study company adopted a Goal Directed Project Management (GDPM) approach, (see reference 2), to overcome this.

GDPM is a team-orientated approach to project planning and management, centred on a process for reaching consensus on what needs to be achieved and who should take responsibility. It breaks down a project 'problem' into milestones which are regarded as measurable states of achievement, not department-based work elements (Figure 4), and is hierarchical, thus avoiding the need for wall-sized PERT networks. Thus GDPM is used as a catalyst to get teams working together, and to demonstrate that vision of project focus.

Making Teams Work

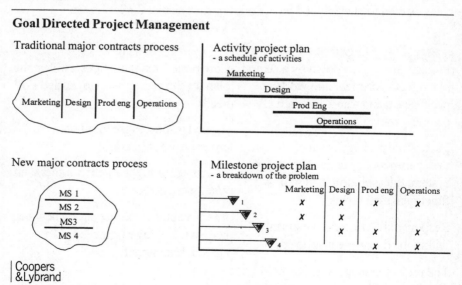

Figure 4   *Goal Directed Project Management*

3.6   Performance Measures

A project business delivering complex multi-disciplined contracts needs a balanced suite of metrics at all levels.   The prime consideration when constructing these with teamworking in mind is always to promote measures which need more than one person to achieve them. Ray Smith, the charismatic Chief Executive of Bell Atlantic understood this, when he said:

> *"There is no such thing as individual success in my organisation, only team success."*

He knew teams were the driving force and the measures he has established are all team-based e.g. salary rises are collective and open, not secret, and recognition of good performance is shared, with no singling out of star individuals.

4 CONCLUSION

Teamwork is not a simple solution to inadequate business performance. All aspects of the system model that is a business organisation must be addressed to access its benefits.

In the company I have described, the teamworking ethic has revolutionised the working lives of many of its people and led directly to a new sense of purpose, projects planned more honestly and delivered reliably and faster, and costs being kept on track. As one project leader said:

> *"This is the most challenging, energising, scary way I have ever worked in 20 years with this company, and it is the only way that gives us a chance of succeeding. It is not an option."*

Significant work is needed to create the environment and potential for good teamworking but it is a necessity in project organisations. There are many shades, but a good team has the creative energy and power to overcome obstacles of history, helplessness, and poor motivation, and enable people to take responsibility to solve their problems. It needs many new skills and supporting frameworks but the rewards are great.

The author is a Principal Consultant with Coopers & Lybrand, specialising in Performance Improvement work with large engineering businesses.

This chapter is based on a presentation given to the Department of Trade and Industry Workshop on 14 June 1995 in Gloucester.

REFERENCES

[1]   A Causal Model of Organisational Performance and Change, by Burke & Litwin, Journal of Management (1992)
[2]   Goal Directed Project Management, by Anderson, Grude, Haug, Kogan Page (1995)

CHAPTER 10

Managing Strategic Process Change

S. Philpott *and* N. Cramer

THOMSON-THORN Missile Electronics Ltd, Hayes, Middlesex

1 INTRODUCTION

In recent years, in common with other industries that design products for manufacture, the Defence Industry has had to actively consider methods of improving their business processes, ahead of competitors who are striving to improve their ability to deliver more products with higher margins. In addition to these pressures, many sectors of the Defence Industry present shrinking markets and competing in this environment means achieving the best in every aspect of business.

Our products are complex and often push at the limits of technology. Generally they are high cost, with long development times, typically 3-5 years. We therefore faced a major opportunity in improving competitiveness by reductions in:

- development cost
- development leadtime
- product costs

THOMSON THORN Missile Electronics (previously THORN EMI Electronics Defence Group) have been active for some time in formalising the design process and introducing Concurrent Engineering philosophy and practices with the aim of achieving efficient, cost effective new product introduction; in short, to:

- achieve customer satisfaction at minimum cost
- optimise timescales at minimum risk

2 THE STARTING POINT

The starting point on our road to improvement, was to use cross functional teams to define, document and introduce a formalised 'Design Process Procedure' based around a typical 'generic' programme. This formal process took more than 1 man year of effort and provided a detailed self assessment process to monitor and review attributes of the design at set stages using checklists. This process was designed to 'ensure' concurrent involvement.

Even in the early stages of this process there were those who realised that this exercise could not be consistently successful without an organisational culture change driven from the top, and it was vital that the Design Process was seen in context with the whole business. The concurrent design methods and philosophy needed to be embedded into the minds of the product design teams: they needed to really believe in it.

From this point THOMSON THORN Missile Electronics embarked on a three pronged attack:

- a pilot Concurrent Engineering programme
- full implementation of the documented design process into the company Business Management System and ISO.
- an awareness programme to engender active support of the concurrent processes

The first two of these have proved successful; the third is an ongoing process of education and culture change which will take a number of years. Within this, the consistent and active direction from senior management will play a vital role, as will ongoing training in the methods and philosophies of Concurrent Engineering.

The following sections identify the main methods, procedures and philosophies that we have implemented in our quest to gain maximum benefit from the process change.

3 THE DESIGN PROCESS - Initial Implementation

The Design Process defined within THOMSON THORN Missile Electronics forms part of a larger business process spanning all requirements for the whole product lifecycle i.e. from customer requirement through design, certification,

THE INTEGRATED BUSINESS PROCESS

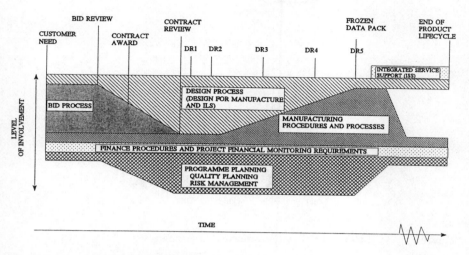

Figure 1   *The Integrated Business Process*

production support and disposal. To depict this, a process phase diagram was generated (Figure 1) showing how the Design Process integrated with other key business functions.

The Design Process can be broken down into a number of key stages; we chose to depict these as a staircase (see Figure 2) with each 'step' equating to a different part of the product lifecycle:

Stage 1 -   Requirement Definition   -   WHAT is required.

Stage 2 -   Product Definition   -   HOW to achieve the requirement - outline concept.

Stage 3 -   Detailed Design   -   HOW to achieve the requirement - detail.

Stage 4 -   Build and Test   -   Prototype build.

Stage 5 -   Qualification & Trials   -   Verify against requirement.

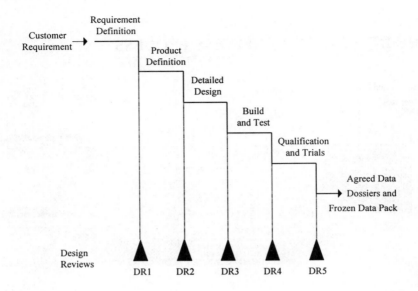

Figure 2   *The Design Process: Activities and Review Stages*

Having identified the critical stages in the process, the cross functional team defined the contents. Each stage was to have:

- A defined process for the design stage
- Defined responsibilities
- Detailed checklist relevant to all stage design aspects
- Stage documentary output
- Formal review with actions

Care was taken during the definition of the process and documentary requirements that they did not strangle projects with unnecessary paperwork, although use on small design projects with <6 months design phase suggests some further alleviations may be required.

Initially the documented process was perceived as a 'stand-alone' solution, but it soon became apparent that it formed only a part of the framework for change and was not the whole. It became clear that the introduction of simply another set of procedures without an accompanying culture change would not work. From practical experience of design teams and the Design Reviews it became evident that designers and management were fighting the process.

Without education, training and on the job facilitation the new philosophy embodied in the documentation was often seen as a constraint to be avoided. Without any central support function to monitor projects day by day, the application of the process in some areas was no more than cursory. Design Reviews became an adversarial process, instead of a useful forum for open self analysis where real improvements could be determined.

In many areas no training was provided to Programme Management or Design Managers in team building and hence individuals did not commit to parallel working. Many elements of the design were still left to be swept up at the end with little direct consideration for product cost and the effect of late design change. It could be said that we had attempted the easy 'paperwork' route to Concurrent Engineering - we now know that this is only a part of the story and the people issues play a major role in achieving success.

Utilising the Design Process as the foundation, further investigations were made into Concurrent Engineering philosophies, tools and practices. This then led to actually working with the individuals and design teams to help them understand their responsibilities and the demands that the new process placed on them.

4 IMPLEMENTING THE CHANGE TOWARDS CONCURRENT ENGINEERING

Having asked the question 'How do we **make** people **want to** practice Concurrent Engineering?' we became involved with investigating the requirements of Concurrent Engineering through industry groups with a number of other companies. We also participated in University research to find out our Company's position with regard to how far we were already using CE methods and where we could improve. It was clear from the difficulties we could see that there was scope for improvement in some design areas - high development costs, late design changes, redesign, and inability to meet the required target costs still persisted.

Traditional methods of design have long been under attack from the concurrent approach for their 'over the wall' mentality i.e. engineering designing, then passing to production (and then redesigning as often as required...). The concurrent approach that we were aiming for (see Figure 3) appeared to be able to give a time saving with a proportional reduction in development cost together with improved confidence in the design solution.

202

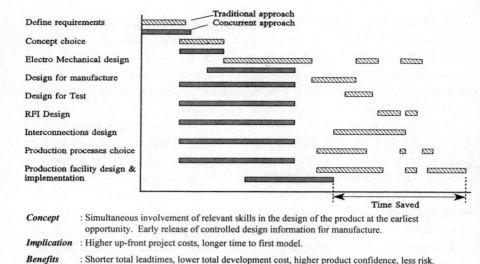

| | Traditional approach |
| | Concurrent approach |

Concept : Simultaneous involvement of relevant skills in the design of the product at the earliest opportunity. Early release of controlled design information for manufacture.

Implication : Higher up-front project costs, longer time to first model.

Benefits : Shorter total leadtimes, lower total development cost, higher product confidence, less risk.

Opportunity : To beat competition by offering lower development costs and shorter development leadtimes whilst maintaining margins. To increase market share by beating competitors to the market.

Figure 3   *The Concurrent Approach*

Experience of companies in similar industries indicate that 10 to 15% reduction in product and development cost may be realistic in the Defence Industry [1]. Some senior managements are still uncertain of these savings, since the non-repetitive nature of our designs, make it very hard to say what costs would have been had we not used Concurrent Engineering! Even our major pilot project will not yield this information, although so far it has succeeded in achieving its target. One advantage is that most of the individual design managers involved in the pilot, although sceptical at first, have been convinced that CE is the way forward.

The realisation that our procedural 'Design Process' was not enough for business improvement was the key to the next step: the cultural change and understanding of the changes of roles and responsibilities in the organisation was critical to focus effort where it was required. The traditional barriers between different parts of the organisation had to be addressed and it was crucial that all areas understood the key factors affecting the successful introduction of a new product.

Figure 4 indicates the key areas of the business which we determined would affect our ability to improve new design introduction, and indicates some of the most important subsidiary factors. The main objectives included:

- reduced development leadtimes
- increased product confidence
- a minimisation of late design changes
- lower product cost.

As part of the implementation process, all of the subsets within Figure 4 were reviewed and their relevant importance to the Company was determined, based on what we believe are the most important for our type of business. Businesses in other market sectors will find other sectors critical to their needs.

As already discussed, procedures and standards are important within our business and so it is not surprising that much of our effort was initially focused on defining and documenting a suitable concurrent process of moving from requirement through product definition to full production. Emphasis was placed on the importance of the procedures within the business and how we worked to them. With hindsight other aspects of Figure 4, such as team needs, facilitation and communication are considered to be crucial areas where the change process is going to be of most benefit to the Company.

New Product Introduction Opportunities

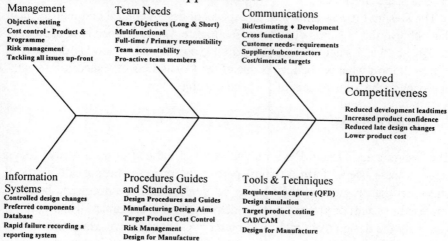

Figure 4   *Key Factors Affecting Successful Design*

The remainder of this Chapter now discusses the principles of our identified 'key factors to success' from Figure 4, together with comments on our successes and failures, with suggestions for potential improvements:

- Procedures Guides and Standards
- Team Needs
- Communication
- Management
- Tools and Techniques
- Information Systems

5 PROCEDURES GUIDES AND STANDARDS - KEY DOCUMENTED PRINCIPLES

5.1 The Documented 'Design Process'

The documented Design Process was generated as a logical and common sense process for the management of design activities. It is defined so that as many problems as possible are identified and solved in parallel, as early as possible in the design cycle. This process is assured through the generation of key documents with defined contents and checklists, and through the Design Process stage checklists.

The Design Process embodies all aspects of design control and management that achieve the business objectives (see Figure 4). The procedural aspects of design i.e. the 5 stage process for activity, output and review as depicted in Figure 2, form what we believe is one of the key aspects of a successful 'team' approach to the achievement of the business objectives.

5.2 Design Review

The design stage Design Reviews that occur at the end of each of the design stage process 'steps' form a critical part of the process and provide a crucial check on the progress of the various activities which are necessary before a successful system or product may be delivered to the customer. Their purpose is to help the design team by providing a critical cooperative examination, first of the design concept, later of its detail and finally of its suitability for production and use.

The Design Reviews act as an assessment of technical progress and ensure that activities have been properly undertaken and adequate documentation has been put in place. They should also aim to highlight any deficiencies in design, by bringing together staff with broad experience of the business to review the project. They form a basic check that the project is 'on track' and that adequate risk management has been undertaken.

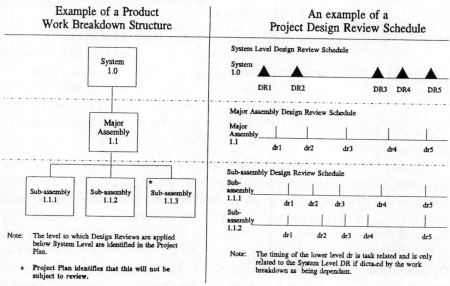

Figure 5   *Work Breakdown Structure and Review Schedule*

The right hand side of Figure 5 shows a typical approach to a Project review plan. The System level design elements are subject to a formal, top level Design Review ('big' DR - see below) at each stage of the design i.e. DR1 to DR5. The sub-system levels are subject to the same Design Process stages but are subject to lower level ('small' dr) review after each stage. These can, and are likely to, run in parallel with both the System level and other lower level design activities (a Project Manager may choose not to impose a lower level review where risk is low). The lower level reviews must be planned by the Project Manager to fit in logically with other higher level reviews as appropriate.

Where there are a number of design iteration phases ending with production hardware, the review sequence ('small' dr 1-5) will be carried out for each item and sub-assembly as appropriate, for each iteration.

- DR — Project level Design Reviews are reviews held to complete one of the 5 Design Process stages during the development of a complete product or system. These will normally involve the appropriate Bid or Project Manager and all relevant support functions. They will have an independent Chairman and may have customer or supplier representation.

- dr — Lower level design reviews can be at the discretion of the Project Manager. Usually they will be held to complete a particular Design Process stage. Generally these are applied to nominated critical sub-assemblies, sub-sub assemblies and components as required. Attendees are usually internal to the project and may involve as few as two individuals, depending on the subject.

The outcome of the formal review was not defined as Pass/Fail but was controlled through constraints and critical actions. A project design could be constrained so that no further progress could be taken until certain actions had been carried out. This provides a powerful external control on a project and if used properly, greatly increases the chances of success.

Initially there was concern by design teams that the Review Process prevented the building of risk reduction test hardware prior to the top level review of the detailed design; this is not the case. It is now stressed to designers that to minimise risk it is necessary to prototype hardware and carry out special design studies that will move into subsequent stages prior to review. The risks of commencing subsequent stages before completion of parallel activities are now clearly identified, monitored and controlled by the programme.

5.3 Design Checklists

For each stage of the Design Process detailed 'product centred' checklists were generated by a cross functional team. The purposes of the checklists are to provide:

- A reminder to the designer of critical activities at the start and during the design stage.
- An end of design stage check that the designer has adequately carried out all activities identified in the checklist.

It is important to get the design teams to use the checklist in the planning of a design stage and to informally review the status at one third and two thirds of the way through the process.

The checklist is used as a review of completion of the activities at the end of the process. Again the use of these checklists is no guarantee that the team will conclude all areas successfully. We believe that the formal documented checklist requirements must be combined with education and training to achieve a positive, pro-active use of these checklists.

5.4 Design Documentation

A list of formal documentation is called for by each stage of the Design Process. In order to minimise unnecessary effort it is stressed that the documents need only be as detailed and complex as their particular area of design requires. If a document is irrelevant then the 'team' may decide that the work is unnecessary; however, the need for the work and the record WILL be reviewed at the stage Design Review. If the work is found to be required then the team will have to go back and complete it.

One of the initial problems was that very few people knew the content of the documents that were being called for, so the interpretation of the need by each design team provided some highly variable output. Manufacturing plans, for instance, ranged from a schedule of customer deliveries, through computer simulation of capacity, to a detailed analysis of the design producibility. To prevent this variability and assist with the education process, guides were generated defining the content of the key documents required by the design process. These guides often contained checklists to assist in the process. Soft copy templates were also found to be useful for speeding up document generation process.

Some of the key 'design for manufacture' related documents are identified below:

- Producibility Plan - How the design team will be supported to achieve producibility.
- Producibility Report - The status of achievement of a producible design.
- Manufacturing Design Aims - What are the constraints that manufacturing applies to the design.

- Manufacturing Plan - What needs to be done to get product into manufacture at the correct, time cost and quality.
- Target Cost Report - Status and action plan to achieve the required product manufacturing costs.

5.5 Manufacturing Design Aims

Generally the end point of a project was a manufactured product and many designers had for some time been requesting a 'design for manufacture' catalogue listing all manufacturing capabilities. This we determined would not only prove impossible due to the interrelationship of the product and process, but would inevitably change as the manufacturing technology progressed during the project. The solution came in two parts, the first was to set the direction in terms of a set of broad 'Manufacturing Design Aims' and the second was to provide the project with a full-time sage and knowledgable manufacturing engineer, to interpret the design aims in relation to the detailed design. The process works as below.

At the start of a project, during the concept phase, the allocated Manufacturing Engineer generates a Manufacturing Design Aims document for the project. This is used to help designers comply with company capabilities and to ensure that past 'design for manufacture' mistakes are not repeated. It defines WHAT we and our sub-contractors want to do in manufacture, not HOW to do it. The Manufacturing Engineer is responsible for assisting with interpretation and implementation of the Manufacturing Design Aims for the design team.

The Design Aims are 'tiered' and the higher 'aim' in the list always takes precedence. Example Design Aims for our business would include:

UPC	-	Better than Target Cost
Electrical Circuit Boards	-	No hand soldering
	-	Surface Mount Technology
	-	No mixed-technology boards
Assembly	-	Use of existing equipment and processes where possible

It is important to note that in the case of conflict the Better than Target Cost always takes precedence - except when this impacts significantly on:

- Design timescales
- Programme cost
- Technical achievement
- Interchangeability / space constraints

These areas usually end with a negotiated settlement with the Programme Manager.

5.6 Design for Manufacture Guide

In order to further improve our ability to Design for Manufacture we generated guides to define HOW staff responsible for design for manufacture should approach a product development project. It covers the responsibility for design input and the documentation requirements. Topics covered include:

- The tasks of the Manufacturing Engineer on a project, enabling a focused review of activities against a specific checklist.
- How a Manufacturing Plan should be generated, who is responsible for generating it and a checklist of contents to ensure that the product can be delivered to the customer requirements with the required profit margins.
- How and when Producibility Plans and Reports should be generated, to demonstrate that planned progress is being achieved for product producibility.

5.7 Target Cost Control

Similar to other industries, an important message to get across to the design team, was that about 80% of our products' manufacturing cost was determined within the design phase (see Figure 6). It was therefore vital that sufficient time and effort be expended within the early stages of the Design Process to monitor and control the product cost. Past history on some projects has followed the scenario in Figure 6, where lack of attention to product cost only becomes a senior management 'hot potato' when it is far to late to do anything about it. These errors clearly demonstrated the need to change the way we designed products. We experienced the problems of additional costs, linked to late design change, as depicted in Figure 7. Since almost all of our design and production

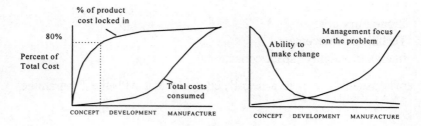

The greatest benefits of cost control are in the Concept and early Development phases.

Figure 6 *Benefits of Cost Control*

contracts are 'fixed price', costs associated with the late design changes and lack of cost focus have, in the past, cancelled out entire product life profit margins.

Unfortunately many designers had never had a product cost responsibility placed upon them. This has been quite a culture shock, but now the responsibility and accountability for cost control lies clearly with each individual Design Engineer. Targets for cost achievement are set at

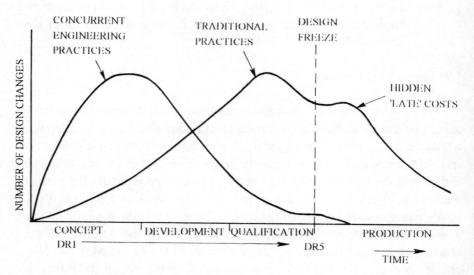

Figure 7 *Effect of Late Design Changes*

the start of the project down to the lowest assembly or sub-assembly for labour and materials, and reviewed regularly to ensure achievement.

Since the Target Cost is fixed at the start, should an individual designer find that their particular design has increased in cost, then they must negotiate within the team for corresponding reductions elsewhere. This process of cost negotiation maintains the original target, but, in addition, brainstorming sessions of design team and 'external' members have been of great value in reducing the total cost, thus increasing profit margins on the fixed price contract.

In our industry Target Costing is one of the most important business controls that can be used in a design team. It should be in place prior to the concept selection phase to enable selection of a concept that meets the business needs. It should be viewed by the design team to be of equal importance to the technical and schedule requirements and should be given equal attention.

Early identification of the Target Cost enables a cost trade off analysis to be carried out, which can assist with early customer discussions of specification requirements against product cost. There are often instances where the customer will not realise the product cost penalty incurred by particular specification requirements and pointing these relationships out is an important factor in retaining future business.

The Target Cost is the first item to be defined so that the bid team can put in a realistic cost. During the early concept investigation this should be the cost basis for limiting concept choice. In our industry the pareto of high cost items at this stage will probably provide sufficient information to determine a product cost compliant solution.

Once the solution has been determined it is necessary to regularly review the product costs. Our experience shows that areas of the design increase rapidly in cost, when working to a specification close to the limits of technology. In order to remind designers of their costs, reports of the CURRENT (average) unit production cost (UPC) of the design need to be generated on a regular basis to compare with the Target Cost. Design changes are only accepted into the CURRENT cost if agreed by the design team that there is a very high confidence of occurrence.

If there is an identified cost saving opportunity or cost increase that requires further investigation then the probability of occurrence of each potential change multiplied by that change provides a forward 'forecast' of where the cost might go to in the future. This we call the FORECAST cost.

No cost reduction can be taken into the CURRENT cost unless an agreed design change has been made. (It remains in the FORECAST until the design change is approved.)

Target Cost Action Plans identifying individual actions for cost reduction opportunities, plus the percentage likelihood of achievement, are generated and regularly updated. Review meetings of outstanding actions, savings, threats and occurrence probabilities maintain the accuracy of the costs.

The FORECAST cost is generated from the CURRENT cost minus the sum of the cost reduction opportunities in the target cost action plan, multiplied by their respective probability of achievement.

Honesty and integrity are important factors in the generation of the costing information. There are numerous pressures that can deflect the costs from reality, and maintaining 'true' costs can prove difficult with mounting cost reduction pressure. It is as important to maintain a focus on the potential cost increasing threats as it is for finding the corresponding cost reducing opportunities. Without this balance a distorted (low) cost will result.

The reporting of the TARGET, CURRENT and FORECAST costs is best done graphically, which provides the opportunity to see trends and drive product costs to a profitable conclusion.

5.8 Risk Management

We have found that one of the key aspects of CE is the avoidance of late design changes due to early unidentified and un-reviewed risks in the early design. The management of the risks will tend to pull forward the decision making process, which causes much more design change activity early in the design and avoids the late production design changes (see Figure 7). The philosophy of 'if it can be done early, do it' is important, since this reduces later risk and will give a cost benefit. For this reason an active documented risk management system is important in the early identification and critical review of risks collected from the whole design team. The high level review of the risks banded by cost, technical implication and probability will prioritise the local objectives of specific design teams.

Management of risk is the responsibility of all those working on a project. All risks associated with the achievements of business objectives must be highlighted to the Project/Risk Manager in terms of

- Time
- Cost (Product and Project)
- Quality
- Technical achievement

The purpose of the risk management process is to identify and analyse the effects of uncertainties on a project and to enable timely action to be taken in terms of time, cost and quality requirements. Risk analysis and management involve **ASSESSMENT** of risk, comprising:

- Identification
- Analysis
- Consequences
- Prioritisation

and **CONTROL**, comprising:

- Planning
- Implementation
- Monitoring

Generally for larger projects there will be a documented risk management plan identifying how the specific project will maintain risk control. A company guide has now been generated, identifying the content and structure of this assessment and control process.

6 TEAM NEEDS

The team is everything. If an effective team can be generated, pulling together for the common objective, rather than operating as a group of individuals, then we believe that 90% of the battle for CE has been won. Generating an effective team is NOT easy, as it demands specialist people skills that often design teams do not have. It will often prove easier for

Design Managers to hold central review meetings than to empower real teamwork, which will progress without their input.

Early efforts in defining the team structure and team objectives for effective working are crucial. Team dynamics and blocking factors are critical to success and require regular review.

The Project Manager needs to be sensitive to team needs and to resolve any issues before they affect team performance. Some examples are:

- Character issues
- Lack of functional resource
- Lack of skills
- Lack of sufficient 'team pro-active' members/facilitators
- Recognition of performance - both individual and 'team'

Other elements that have been found to be important to help the team to function effectively include:

- Core team having full-time members.
- A number of pro-active team members who are able and willing to coordinate across function boundaries and to maintain communication links.
- Part-time specialists e.g. RFI, stress analysts, as required by the core team - but kept in touch via updates of project progress.
- A multi-functional core team, made up of key project staff (e.g. Design Engineers, Manufacturing Engineers, Procurement staff etc.), the content of this team being dependent upon the stage of design and the project type.
- Clear and regularly updated long term and short term objectives for the team and for individuals.
- A Project/Design Area Manager as an active core team member.
- Clearly allocated team and individual responsibilities for work topics and areas.
- Authority of the core team to make design decisions.
- Core team accountability and responsibility for design decisions.
- Key sub-contractors considered as an extension to our core design team to ensure concurrent design.

7 COMMUNICATIONS

Communication links at all stages of design are crucial to the achievement of overall business objectives. It is vital that information can flow in the right directions, between the right functional areas and between design teams and suppliers/sub-contractors and our customers.

The new computer interactive EDM/PDM (Engineering Data Management/Product Data Management) solutions can be seen as easy 'no effort' ways of solving what is essentially a human interaction problem, involving trust and face to face negotiation. It is doubtful that, if the team is not working effectively in the first place, the imposition of such a system will have any effect. These systems will, however, provide much better information management and rapid access to design and change data which will considerably enhance the effectiveness of a working team.

Key areas of communication that have been found to be most effective include:

- Open and honest two way communication with the customer (both internal and external), to ensure requirements are understood and risks identified.
- Open and honest two way communication between the team members, supporting functions and senior management.
- Open and honest two way communication at internal reviews and with senior management.
- Good communication with the customer at the bid and estimating stage; this may require a requirements capture tool e.g. QFD - Quality Function Deployment, as an aid to communication.
- Close links between 'in-house' project contributors and suppliers / sub-contractors. A sub-contractor may even be part of the core team.
- Use of CAD solid model as a team focal communication point for early concept design.
- Use of Circuit Simulation Tools such as SPICE^{TM} and MENTOR^{TM} for analysis of electronic performance.
- Good communication between marketing and potential customers for future product needs and potential sales (e.g. mid-

life upgrades, product flexibility etc.)
- Use of management communication tools such as 'brainstorming' to attack specific problem areas.
- Use of a PDM tool enabling common access to product data and assisting with the change control processes.

8 MANAGEMENT

There are many 'Management factors' which affect the ability of a project to operate in an environment of improved competitiveness - 'Management' in this case is not project management, it covers all those managing staff or interacting in teams.

It is important for objectives to be set at the start of the project or at a particular stage of a project - these will include key 'milestones' or goals to be reached at particular points through the product lifecycle. Regular reviews of these global or local objectives will ensure the directed efforts of individuals or sub-groups to the project's common aims.

A concerted approach by management to tackle all issues up-front in the project, with a general project philosophy of 'tackling all problems as early as is sensible within the design cycle'. This will be key to facilitating 'right first time' manufacture and achievement of the principles of Concurrent Engineering.

Late design input has been proven in our company to cause higher design and production costs and technical risk. The general idea that "if I ignore it, it may go away", is very dangerous because it always comes back! Unfortunately with a design cycle extending over a number of years, members of the 'team' can move to other areas within the business before the manufacturing phase is entered. Accountability and responsibility for design problems can thus be diluted, making it easier to criticise manufacture for high cost and manufacturing problems related to design issues which had not been resolved (see Figure 7).

It is the responsibility of all involved in the project to identify areas of weakness and encourage (or coerce!) until the required design input is made. This is where the pro-active facilitator can be most beneficial as a member of the team.

Another important management factor is the consistent approach to the management of risks (see section 5.8). An important function of the design managers is to identify, document and circulate risks to those responsible. Also a critical role of the Design Manager is the encouragement and assistance to individuals to identify and control their real risks.

The traditional project planning/management tools (PERT) play an important part in the resourcing, tracking and objective setting of the teams. These systems are vital on the larger projects and we have found that Concurrent Engineering enhances the ability to achieve shorter timescales but does not replace the planning and tracking tools. A good integrated and locally accessible planning tool provides visibility to the design teams of their local design/development cost and timescale targets, and can provide the team with local objectives.

It is important to remember that even with Concurrent Engineering, the type of product will determine the extent of team support through into production. Our products are at the 'leading edge' of technology and so we generally need support disciplines to be involved for the first 6 months of production.

9 INFORMATION SYSTEMS

We believe the key information areas to support concurrent working are:

- Rapid, easy access to current design information.
- Preferred Parts/Component Database. A choice of common parts or components with current projects, giving less handling, issuing, space, purchase orders and verification at source (VAS) activities, with an associated reduction in the amount of 'control' data to be managed.
- Approved Suppliers List, giving a reduced number of suppliers, leading to a reduced number of orders, VAS management activities, communication chains etc.
- Rapid failure recording, reporting and solving, ensuring that an individual takes 'closed loop' control for any design or product problem. This should ensure rapid implementation of:
    - Corrective action (short term fix).
    - Preventive action (long term solution).
- Simple and controlled product configuration

- Direct links to both CAD/CAM, Database and Manufacturing control systems (MRP)
- Utilisation of cost control for on-line spend data.

These requirements point to the use of an integrated EDM/PDM system that can provide rapid access to accurate design status information. The system should enhance our ability to make the design change process efficient and aid cross functional communication.

Our implementation of an EDM system has proved reasonably successful. In reality the system was designed mainly as a configuration control tool and not as a communications enhancing information system. There is much more in the way of PDM available 'off the shelf' than when we started our implementation.

10 TOOLS AND TECHNIQUES

Although there are many tools and techniques that impact on CE, we have identified below the key ones which we feel have enhanced our Concurrent Engineering effectiveness:

- Requirements capture (Quality Function Deployment - QFD) - this results in the design being specified more fully earlier than would normally be seen in the process of design. This provides greater confidence in the design with a focus on the critical factors.
- CAD/CAM/CAE - including proper use of an efficient CAD/CAM system as a design communication tool and as part of the Engineering Data Management System.
- Design Simulation - use of simulation to enhance early design confidence.
- Rapid prototyping - Stereolithography - use of a CAD generated solid model to generate model hardware directly to provide early hardware confidence.

11 THE WAY FORWARD

In conclusion we believe that the success of Concurrent Engineering in any company will be a combination of factors (Figure 4). However, the relative importance of these to a company, the order in which they are implemented and the level of success achieved, will depend on many factors:

- Business sector - external environment
- Company culture - internal environment
- Skill level and understanding of the company - starting point & need to change
- Level of understanding, ability and focus of the directors - drive and measures for change
- Commitment to training and personnel development

For us, by far the most important factor in changing how the Company works has been the degree of acceptance of a major cultural change. We started with the documented definition of the design process and then moved into the area of teamworking and Target Costing. It is difficult to place success in any particular area but in our opinion the successful combination for us has been:

- 30% Documented processes, procedures and guides
- 40% Teamworking, team needs and communications
- 20% Management factors
- 10% Information systems and others

We have found that to be successful the change process has to be driven from the top (and needs to be **seen** to be driven from the top). This requires a good level of understanding and drive from the directors to maintain achievements.

Good communication links have needed to be forged between all areas of the business (in particular between those 'factions' who have traditionally been at odds with one another).

We have found that it is important that the traditional 'apportionment of blame' mentality is tackled and removed from the 'teamwork' environment - all team members need to be aware of (and accept) their joint responsibility for project issues. Strong two way interaction is critical in removing the 'us and them' philosophy.

Teamworking has cut across the traditional boundaries and can challenge the decision making process and the authority of the baronial structure [2]. Many managers have found it difficult to cross the divide and lose their traditional functional control system to more modern measures of project success. There has been a tendency to treat the new ideas with suspicion and many have been very quick to see short term problems as a sign that the system has failed.

We have realised that the process of changing culture and moving towards a new way of working takes time. With a design and development cycle of 3 to 4 years, proving the point to a sceptical management team can be very difficult.

Staff issues in terms of training in the new ways of working, team responsibility and improved communication still need to be actively addressed. However, by demonstrating the success of the pilot CE project there is now commitment to change **throughout** the Company. The pilot has now provided a nucleus of 'converted' supportive design team members and Design Managers, who will be able to spread the word onto other product design teams.

As a Company, we have used the documented Design Process and Concurrent Engineering philosophy to go a long way towards improvement. With hindsight, in the early stages we may have spent too much effort on our procedures and now recognise the importance of team needs and communication as being vital to the change process. On the other hand, the process has to start somewhere and our success is that we actually have started and that we believe we know where we are going.

12 REFERENCES

[1]   Barton, R., Siemens Plessey Systems, IEE Colloquium "Current developments in concurrent engineering methods and tools" 2 June 1994.

[2]   Constable, G., Concurrent Engineering - its procedures and its pitfalls, IEE Engineering Management Journal, Vol 3 No. 5 Oct. 1993.

PART IV

BUSINESS PROCESS RE-ENGINEERING AND IMPLEMENTING CONCURRENT ENGINEERING

CHAPTER 11

Implementing a Business Process Re-engineering Programme

J. Paul *and* C. Burns

1 INTRODUCTION

BTR plc is a major global engineering and manufacturing business employing 120,000 people worldwide, with sales in 1995 of £9.5bn. Core businesses include Power Drives, Process Control, Automotive Components, Packaging, Polymers, Building Products and Specialist Engineering.

The acquisition of Hawker Siddeley by BTR plc in 1990, introduced new management and new thinking at a point in time when the need for a quantum change in direction for BTR was recognised by both internal and external audiences.

The period of introspection which followed on from the integration of the Hawker businesses allowed some degree of freedom, not previously available to BTR management more used to working within a tough financial control culture.

For BTR's Aerospace Group, the timing was opportune and a formal strategic planning process was designed and applied in selected businesses. Also at this point, Business Re-engineering, although at an early stage in one or two of the Hawker aerospace businesses, was sufficiently advanced as a body of experience to form the base for change programmes for more businesses, as they completed their strategic reviews.

Following early success, in an environment in which rigorous financial measurement ensured that performance improvements were real and substantial, the strategy development process was applied more broadly across all product groups leading to increasing demand for strategy driven BPR programmes.

The next step was the appointment of a Change Manager, reporting to an

Executive Director. Principles and processes were developed which essentially provide a framework for parenting, team based, bottom-up driven change, with appropriate central support.

The combination of a powerful strategy development process with the enabling tools of BPR has created some exceptional and lasting performance benefits for individual Strategic Business Units (SBU's). This paper attempts to outline the basic processes and principles which are at work in BTR today.

1.1 What Does Business Process Re-engineering Mean to BTR?

BPR requires a fundamental change in the way a business does things and sometimes it requires major changes in the things that it does. It requires commitment to a new way of life, and by its nature is long term. It is not a quick fix.

BPR can be defined as the fundamental analysis and radical redesign of the whole business, including:-.

Business processes
Management systems
Job definitions
Organisational structure
Beliefs and behaviours

The ultimate objective is the achievement of significant improvements in meeting customer and shareholder needs by harnessing the full commitment and capability of our work forces.

The focus should principally be on methodology (the way we do things) and the identification of appropriate areas of investment in technology, which will improve performance further and act as an enabler for process changes.

BPR requires the identification and design of 'value-added' activities which are attractive to present or future customers. Any other activities (non value-added) are designed out or minimalised. Natural Groups of people and equipment can then be positioned around the process, allowing the value to flow to the customer.

The application of BPR principles in BTR has been gaining momentum over the past four years and successful implementations have been used as the comparators and benchmarks to encourage further projects elsewhere in the company.

Reductions in manufacturing and product introduction lead times of between

50-75% are typical, and productivity and value-added per employee have increased by as much as 25% and 30%, respectively.

These and other process improvements have enabled the businesses concerned to gain significant increases in market share, profit growth and return on investment.

Our success in BPR has come not just from a keen understanding of Business Processes and Change Management, but also from a strong belief in the need for any business to have a clear strategy before embarking on a process redesign programme. The strategy must, therefore, provide clear direction and focus for investment of both time and money.

The rest of this chapter deals with the issue of formulating a BPR programme that is fully integrated with the strategic planning process.

2 STRATEGIC PLANNING

One of BTR's values is that management must have ownership and be accountable for their operations. This is particularly important for the business

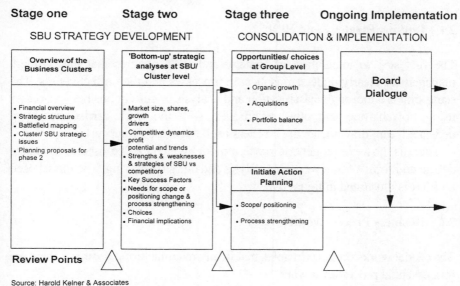

Figure 1   *BTR Cluster / SBU Strategic Planning Process*

unit strategic planning process which has been designed to develop in a management team a sense of purpose and direction.

The strategy process adopted as a pilot in 1995 is illustrated in figure one. The final outputs of the process are two sets of action plans which identify how objectives are to be met. These plans can be broadly considered under the two headings of Scope and Positioning plans and Business Process Strengthening plans.

Scope and Positioning plans address the issues of:
> What business should we be in?
> Where should we do business?
> Which customers should we serve?
> Via what channels?
> Why should we be in this business?

Process Strengthening plans address the issues of the internal strategy:
> How should we do business?
> - Which business process can/ must be strengthened?
> - Which processes should be re-engineered?
> - What is needed to ensure ownership and implementation?

2.1 Linking Strategy and BPR

The first step to integration of strategy and BPR requires that the business management clearly understands its core processes and their effectiveness. The start point is a thorough classification and analysis of current business processes using flowcharting and other techniques, with the work carried out by a dedicated team drawn from the business itself to ensure ownership.

There is, however, a generic process model which can be used to stimulate debate and form a framework for analysis and decision making. The model used by BTR is illustrated in figure two.

2.2 Business Process Model

The model works on the principles that in any manufacturing business, there are four essential processes at work.

The Development Process is essentially the "tomorrow" process i.e. all those things that a business must do to ensure success for itself and its stakeholders in

Business Process Model

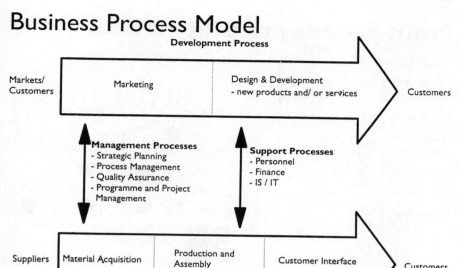

Figure 2   *Generic Business Process Model*

the future. This process takes inputs from the market and develops new products or services for customers.

The Operations Process is the "today" process i.e. all those things that have to be done to achieve success today. This process takes inputs from suppliers, adds value and distributes to the customer.

The Support Process takes inputs from the other business processes and provides support services.

The Management Process recognises requirements of the stakeholders and regulatory bodies and, having compared these with all the internal process outputs, formulates policy and direction for the business.

2.3   Developing a Process Strengthening Plan

Once the strategy is formulated and the business has a clear view of its key business processes, then careful analysis and consideration of options and choices can be carried out. A programme of change is then constructed, and figure three illustrates the approach.

From Strategy to BPR

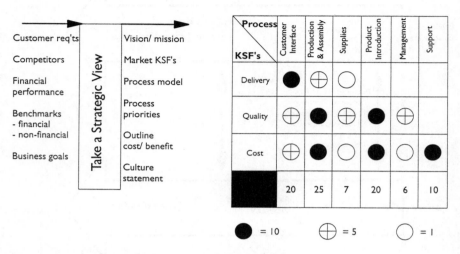

Figure 3 *Linking of Strategic Outputs and Business Processes*

Key Success Factors derived from Customer Selection Criteria are referenced to the major business processes and weighted according to the impact that the particular business process is assessed to have on the KSF.

A simple addition then enables the business to rank processes in order of impact and, therefore, priority.

The next stage is to consider the plans for Scope and Positioning, bearing in mind that they will almost certainly be dependent upon the effectiveness and efficiency of either individual or groups of business processes. For example, a need to introduce a broader product range will have close links with a Development Process and plans to open up a new sales territory with warehousing depots will require close attention to Customer Interface and Materials Management Processes.

Careful consideration of Scope and Positioning issues as they relate to business processes may result in a revised prioritisation of plans for Process Strengthening. Indeed it is our view that BPR programmes often get bad press when management get too focused on internal issues and lose sight of the vital issues of customer satisfaction and competitor response. Given the many successful implementations of BPR within BTR, we are convinced that these failures are due to the management's lack of attention to prioritisation of the real strategic issues, given finite resources.

A typical example seen elsewhere is the use of BPR tools aimed at the narrow objective of cost cutting or "re-engineering", in common parlance, rather than towards broader strategic aims.

2.4 Delivery Process

Earlier in the chapter we described BPR as the fundamental analysis and radical redesign of the whole business. Inevitably the thought of undertaking such a task will send some General Managers running for cover. BPR cannot be forced on people in a top-down manner. For BPR to work the management team must believe in and have ownership of the process. BPR cannot be owned or driven from a management centre that is remote to the business.

The role of the centre is to influence the business through logic and debate, and it must be seen to be fair and objective in its approach. There are essentially two issues which the centre should seek to influence.

1) Management's "perceived need for change"
2) Management's "capacity/ability to change"

Role of Change Management

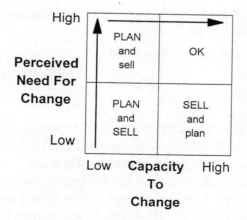

Figure 4   *Understanding What to Influence*

Failure to achieve a common understanding here between SBU management and centre may result in at best a false start, but in some cases this may jeopardise the whole change process.

Whilst visits to re-engineered sites, case studies or diagnostics may help in achieving a common understanding of the potential, we recognise in BTR that real buy-in can best be achieved when centre support is demonstrated by a risk free start up for the management team.

Essentially this comes down to a risk sharing approach between the Centre and the business, which is formalised in two major phases: *Discovery and Execution.*

Change Process

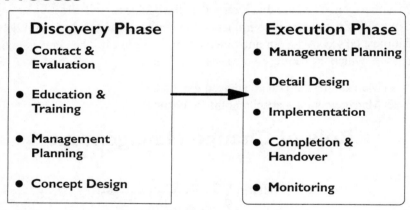

Discovery Phase	Execution Phase
• Contact & Evaluation	• Management Planning
• Education & Training	• Detail Design
• Management Planning	• Implementation
• Concept Design	• Completion & Handover
	• Monitoring

Figure 5   *BTR Change Management Process*

2.5 Discovery

The Discovery Phase allows the business to undertake a process of learning without commitment to delivery of any benefits to BTR, by the development of a Concept Design for the business, with associated plans.

Initial discussions with Senior BTR Management would lead to a meeting with the BTR Change Manager. The role of the BTR Change Manager is to guide the business through the change process and to provide education, advice, facilitation and diagnostic skills.

Workshops, training and visits to benchmark sites would then follow and lead to the eventual selection of an external or internal consultant who will work with the business during the development of its Concept Design.

The Concept Design is carried out over several weeks against the background of the strategic plan. The Concept Design further prompts the SBU managers to consider a realistic vision for the future and delivers detailed plans on how to achieve the strategic goals identified. The process typically asks the question, "if we were to apply best practice BPR techniques and principles to this business, what would it look like?" i.e.

How can we maximise value-added, as perceived by our customers, and
   eliminate NVA?
How can we maximise ownership by the work force?
What are the business processes and natural groupings?
What would the organisation structure look like?
What would the roles and responsibilities of management be?
How would we measure the business?
What type of control systems would we use in the processes?
How would we layout the factory floor and offices?
What targets are we aiming for?
How much will it all cost?
What are the benefits?

At the end of the Discovery Phase a decision has to be made by the business and BTR as to the merits of continuing and implementing the new Concept Design for the business. Of crucial importance will be external issues such as market threats or opportunities, but several internal factors can also affect this decision. For example, the Managing Director of the business may feel that some of his/her direct reports are not suited to the new job specifications and may need time to work on replacing or retraining them. Another influencing factor may be that the Managing Director has not displayed strong leadership qualities during the Discovery Phase, and the Centre may then feel uncomfortable continuing. Timing will also be an issue to consider, especially with respect to funding.

More often than not, however, the decision to proceed is given and the detailed documentation is prepared, identifying all costs and benefits and submitted for signatures to ensure full 'buy-in' at all levels. The formality of this step is important in crystallising responsibility and quantifying the financial case for the new business design.

The mechanism for delivering the Concept Design is usually a full-time multi-disciplinary task force, drawn from the SBU and supported by the selected consultancy and the BTR Change Management Group.

2.6 Execution

The vision is turned into reality in the Execution Phase. A plan would consist of design and implementation work packages for each business process identified for attention. Figure 6 illustrates a typical execution plan.

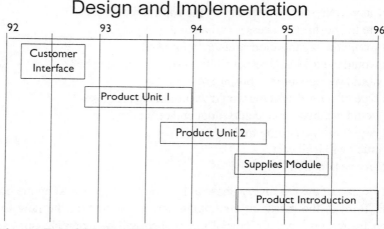

Figure 6   *Typical Execution Plan*

Each project would be undertaken by a multi-disciplinary task force consisting of full-time and part-time people. The team would be supported by a full-time consultant and the BTR Change Management Group would provide technical and change management support to the management team.

The design stage has two elements: Steady State Design and Dynamic Design. Steady State Design assumes perfect conditions e.g. no variations in customer schedules, 100% on-time delivery by suppliers, no absenteeism, whereas Dynamic Design then adds in the variables and uncertainties of life, so that the appropriate planning and control systems can be designed. What this approach achieves, is to ensure that management understands the impact of variation on process measures of performance. Planning and control in isolation cannot account for and deal with all forms of variation typical in business processes. Elimination of the root causes of variability is the preferred choice.

Process design should aim for simplicity, and simplicity will give ease of use and ease of use encourages flexibility, forming the basis for empowerment. A key objective of BPR is to generate an organisation design which is user friendly for everyone who interacts with that organisation.

In many of our re-engineered factories Material Requirements Planning 2 (MRP2) systems have been replaced by what has been termed our "plywood" IBMs. These are large, colourful control boards which are used by the process teams to plan, control and monitor every aspect of their work, with visibility being the watchword.

Careful planning is needed for a re-engineering programme and a great deal of thought has to be given to appropriate Change Management Processes such as communications and project management.

2.7 Project Management

Project management is a key management process which ensures that change happens as planned and in a way which meet the needs of the business. This is accomplished by employing best practice in three areas: planning, monitoring and control.

Planning
Best practice planning should ensure that change takes place with the minimum risk and that resource utilisation is optimised and project execution holds no surprises. Careful and practical resource planning, especially of scarce skills, is often the key.

Monitoring
The execution of the plan needs to be closely monitored and ownership of senior management needs to be visible and regular. Best practice will flag deviations from the plan at the earliest time, ensuring that minimum controlling and corrective actions are necessary.

Control
Controlling actions are used to bring a project back on track and may result in deviations from plan. Choice of the controlling actions should be based on a prioritised list of objectives.

There are many issues to consider in the management of change within a business and there are many common factors which can cause projects to fail. Careful attention to these factors is essential; therefore, within our BPR projects we would always consider the following to be critical for success.

- Full-time Project Manager and task force
- Agreed budget with project manager
- Multi-disciplinary input to ensure coordination between specialist functions
- Discourage inter-function politics
- Have regular, visible, formal planning, control and action review mechanisms
- Keep a focus on meeting milestones
- Install a project specification change approval mechanism
- Have clearly documented project terms of reference and objectives. Specify deliverables if necessary
- Monitor project costs/benefits and plan on a regular basis (typically monthly)
- Ensure the task force have adequate briefing and training
- Insist upon a good balance of formal and informal communication processes

2.8 People and Leadership

With strong and committed leadership a step change in performance can be achieved through BPR. If you do not believe that you have adequate leadership within the SBU, then we would advise against starting the change process until adequate leadership is established.

3 SUMMARY AND CONCLUSIONS

To succeed, we believe that a BPR programme is best designed within the framework of a well developed strategic plan. In the BTR case described, those plans will have been constructed by the SBU management team via a formal strategy development process, with formal face-to-face reviews at pre-determined stages of the strategy process, at Executive Director level.

Without clear commitment to the change process from management and the bulk of the workforce, the programme is unlikely to succeed. Effort spent in communicating the need for change and developing a common understanding of current and planned business processes and activities is vital. Openness with the workforce about change is essential and 'rumour' boards, open days in the task force workroom and change broadsheets are examples of the open management approach. In short, communicate as never before. Typically our redesigned businesses will have only one or two levels of management and these flat organisations greatly facilitate communications.

Budgeting commitment to resource and cost is another essential feature. In BTR, businesses are encouraged to build BPR plans into their annual profit plans along with other strategic activities.

Part-time task forces made up from non-essential or sidelined managers are a big mistake. Pick the best people and plan for full-time commitment whenever possible, usually by back filling their roles in the organisation.

BPR programmes which are dedicated and fully supported will quickly produce a step change in performance. Maintaining and building on the gains requires careful management, appropriate tools and a continuous improvement culture. Planning for this continuous improvement is essential from early in the process.

If BPR projects are too narrowly defined, the results will naturally be limited, however, running too many projects in parallel introduces additional risks and it is better to sequence discrete projects.

Management do need to understand in advance that a BPR programme properly executed is likely to place enormous pressures on them individually. This will not be an 'easy' process and the transition for example, from a management control culture to team based leadership, may result in casualties.

Avoidance of grand project banners and slogans is our preferred approach. If management want to create a visible sign of change spend money improving facilities for employees. Fixing up the rest rooms can send a stronger message than a fancy slogan.

Setting clear stretch targets that have been fully discussed and agreed is a key to success and commitment to these by the management team is essential.

In conclusion, our message to the business teams can be summarised as -

Think Customer Value
Think Radical
Think Lateral
Think Structure
Keep Your Nerve

Finally, keep in mind that the combined processes of strategy development integrated with BPR should be designed to address two key audiences - the customers and the work force. Bringing these two groups onto the common ground of needs, value and performance is a powerful and attractive solution to creating value for a third audience - the shareholder.

CHAPTER 12

Developing a Strategy for Business Process Re-engineering

S. Jackson

INTRODUCTION

The Post Office introduced Total Quality in 1989, which led to identification and action on a number of improvement opportunities at all levels. Benchmarking with Baldrige Award winners in 1990 and 1991 led to significant/major changes to a number of processes e.g. planning, organisation structures and people issues such as the introduction of a leadership charter and 180 degree feedback.

However, unlike many other organisations, the Post Office did not initially use Total Quality to target and improve processes. Identification and improvement of processes did not commence until 1992 with the introduction of mapping for all Business Processes and the WesTip improvement methodology (bought under licence from Westinghouse). The focus to date has largely been on continuous improvement and incremental change, through the mapping of and introduction of measurement systems into the key Business Processes. This, alongside the introduction of the EFQM (European Foundation for Quality Management) Business Excellence Quality model and self-assessment against it, have been the priorities for the three main Businesses within the Post Office.

The Post Office is split into three major businesses (see Figure 1): Royal Mail, Parcelforce and Post Office Counters Ltd. Purchasing + Logistics Services is one of a number of Group Services who operate on behalf of all the Businesses and charge out for their services. It is in these smaller units that there has been interest and action in Business Process Re-engineering and Process Management.

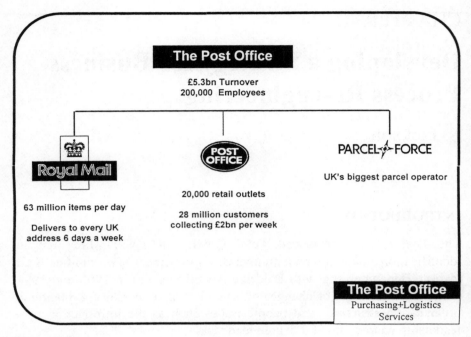

Figure 1

Purchasing + Logistics Services has three key roles:

- – Strategic purchasing
- – Central warehousing
- – Purchasing compliance (EC regulations).

Until the arrival of a new Director at Purchasing + Logistics Services in 1993, the Post Office had had little interest in purchasing and supply, seeing it as just another support function. There was no real understanding of its importance or the contribution to profit that it could make. The new Director instigated a diagnostic of all the Post Office supply chains to understand the size, scale and improvement opportunities available. The diagnostic was the beginning of a major review which became the key platform for future Purchasing and Supply strategy and it is still going through the implementation stage of the actions identified.

The elements of the supply chain were identified as:

- Demand Forecasting
- Purchasing
- Contracts
- Inventory Control
- Administration
- Warehousing
- Physical Distribution
- Waste Disposal and recycling
- Financial Interfaces

The size and scale of the supply chain identified in the diagnostic was somewhat surprising bearing in mind that the Post Office is a service organisation.

- Bought-in costs   £1.5 bn
- Suppliers   6,500
- Operating costs   £43.6m
- Inventory   £38.8m
- Stockholding points   1144
- Staff   1089

As well as identifying the size and scale of the supply chain, the diagnostic also identified what the customers thought of it, which on the whole was not very complimentary:

- Internally focused
- Duplication of activity
- Local sub-optimisation of cost
- Poor communication
- Poor transaction control
- Some islands of excellence, notably the secure stock pipeline and break bulk for print

The Post Office purchases and/or supplies a very wide range of goods and services and the best way to manage them appeared to be to group them into product areas:

- Clothing/uniforms
- Automation/engineering
- Vehicles/fuel
- Print/stationery
- IT/Computers/Office machinery
- Facilities management
- Freight services (carriage of mail)
- Professional services (consultancy, advertising, manpower etc.)
- Value stock (stamps, postal orders)

PROCESS MANAGEMENT

A considerable amount of work was done from 1994 to 1996 to introduce and run Product Group Teams (PGT) as the first cross business/cross functional teams in the Post Office which allowed Purchasing + Logistics Services to manage purchasing as a process rather than as a function.

The teams consist of people with a specific interest in the product or service e.g.

- Commercial Manager - full-time purchasing professional
- PGT Director - a senior director from one of the Businesses with a particular interest/expertise in the product/service
- Specifiers - people who write the product/service specifications, again from the Businesses
- Budget Holders - for the purchase of the products/services
- Major Users/Customers - for the product/service

The key tasks of the Product Group Teams are to:

- Develop and ensure implementation of purchasing, acquisition and supply strategies for their product/service
- Oversee the supplier accreditation and management programme
- Work with the Businesses to help them achieve their targets/goals (see Figure 2)
  - Increase profitability   }purchasing synergy, economies of scale
  - Reduce costs   }common business specifications
  - Increase customer   }brand/image of print, buildings
    satisfaction   uniforms

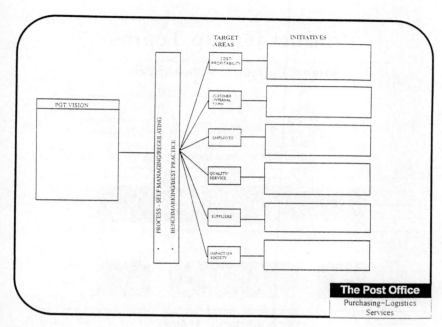

Figure 2

- Increase employee satisfaction   }uniforms, training
- Supplier management   }optimise numbers, performance measurement and development
- Impact on society   }environmental purchasing, disposal, recycling
- Quality of service (Royal Mail/ Parcelforce)   }sorting automation, provision of vehicles and carriage for the mail
- Process improvements   }better, faster, cheaper ways of working

The teams meet 4-6 times per year to determine high level plans and review progress against them. The actual work is performed by sub-groups led by members of the PGT. These are all part-time groups who come together to perform a task/project and then disband; the work is additional to their day-job. The only full-time group is the buying team led by the Commercial Manager (see Figures 3, 4 and 5).

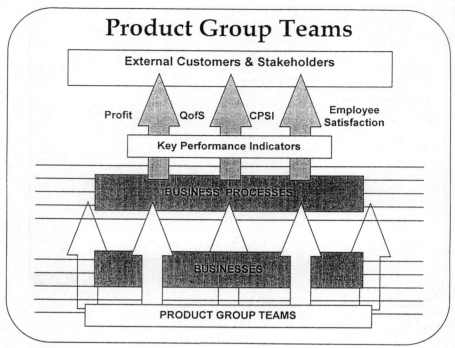

Figure 3

Figure 4

PGTS: MANAGING PURCHASING AS A PROCESS

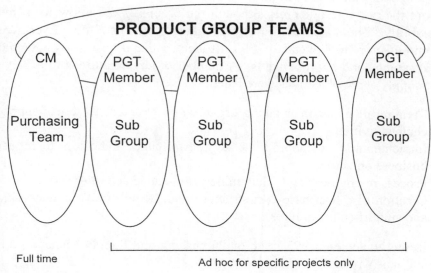

Figure 5

To date, the PGTs have been extremely successful, raising their profit contribution from £9million to £25million in their first year of operation and sustaining this level of contribution in their second year. Thousands of pounds have also been saved due to process improvements which have identified duplication, wastage and new ways of working. Best practice within the different teams is now captured and shared for introduction across all the teams who are currently developing product strategies for the future.

BUSINESS PROCESS RE-ENGINEERING

While the PGTs were being set up, the Supply Chain review continued to work on identifying the future role for the supply chain in the Post Office. The review team had representatives from all the Businesses and Business Units within the Post Office. They examined the bought-in costs of goods and services and located where the supply chain, in particular Purchasing + Logistics Services, could add value.

At the same time, the Government was reviewing the Post Office with a view to privatisation, but there was no attempt to make any judgements about the possible outcome, although the proposals did allow for some flexibility. The diagnostic report was taken to the Post Office Executive Committee on 30 November 1993. It identified and quantified potential cost savings and improvements. A number of key conditions to achieve the supply chain objectives were listed:

- Creation of a customer facing organisation structure to reflect supply chain activities
- Implementation of appropriate key performance indicators to reflect business objectives
- Process re-engineering to determine value-added activities
- Creation of a lean base cost structure with demonstrable added-value and simplified procedures

In the November 1993 issue of Purchasing and Supply Management, Eric Evans wrote:

"Business Process Redesign (BPR) has emerged as management has recognised a need for continued and often radical improvements. There are an increasing number of organisations which have looked at the whole of the purchasing process and redesigned significant elements of it. Their experience suggests that others should consider BPR. The key is that it provides an opportunity to review the purchasing process rather than the purchasing function. It is this broader perspective which generates the benefits."

The Post Office Executive Team endorsed the recommendations, which allowed the Director of Purchasing + Logistics Services to make major changes to his executive team by introducing change agents who would be responsible for implementing the large scale changes which would need to take place. The only areas not touched were finance and personnel. The new posts were:

- Purchasing Director
- Operations Director
- Strategy and Projects Director
- Quality and Business Process Director

With the appointment of the Phase I project leader to the Strategy and Projects Director post, there was a requirement to look for a new project leader for Phase II, the planning phase which was due to report in June 1994. The posts in the executive team and for the project leader all took far longer to fill than originally anticipated and this had a significant impact on the project.

Phase I had been successful but it was agreed that there were some areas which could have been done better or been better managed e.g. identification of the resources required; identification of the tasks and critical path, and a plan to ensure the end date was met. To improve this, a training course for the key players was run on PRINCE project methodology. Few of the people involved had had any project management training, and with the introduction by the Information Technology department of the PRINCE methodology, it seemed appropriate to use the same as systems would play a major role in the supply chain changes.

PRINCE is an acronym for PRojects IN a Controlled Environment. It is owned by the CCTA (Central Computing and Telecommunications Agency) and evolved in the UK government sectors in the 1970s. It is now a publicly available open product. There are a number of principles in the methodology:

- Organisation
- Planning
- Controls
- Stages
- Product based projects
- Quality

The overall objective is to ensure that the right people make the right decision at the right time.

Despite the recommendations being endorsed, the key players trained in project management, and the setting up of a project support office, the project could not move forward for several months because no project leader had been appointed. This was not through lack of trying. Organisations are advised to use the very best people on BPR projects; unfortunately they are not always willing or able to participate. The project hit both these problems and finally filled the post in late February 1994. As there were a number of individual projects already in force within Purchasing + Logistics Services, these were brought together to ensure integration into the overall strategy. These included the First Class Supplier Programme (the supplier accreditation and management programme), the Product Group Teams, and new areas such as change management, communications, business process re-engineering and the development of the business case to secure funding for the implementation of the planned actions. The scale of the projects, however, indicated that the programme would become too unwieldy and it had to be split between the project leader and the Strategy and Projects Director (see Figure 6).

- Major individual projects
- Integrated projects

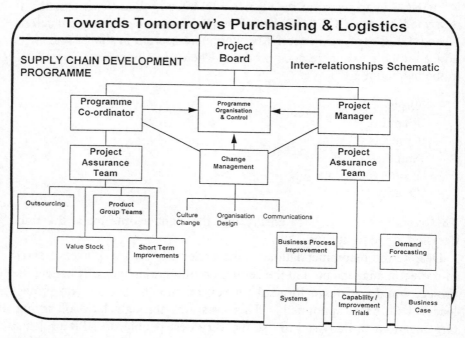

Figure 6

The Quality and Business Process Director was asked to lead the re-engineering project. Setting our sights high, we decided to re-engineer the whole supply chain from the suppliers, all the way through the organisation to the external customers. The projects in Phase II were already operating to tight timescales due to the late start, so there was tremendous pressure to get started since other projects were dependent on the outcome of the re-engineering. What became clear very quickly, was the absence of a methodology for BPR. In 1994, only a couple of books had been published and these mostly dealt with the benefits to be gained from introducing BPR rather than explaining how to go about it. A methodology was finally offered by the consultancy advising on the supply chain review and this was used. This methodology is outlined in Figure 7.

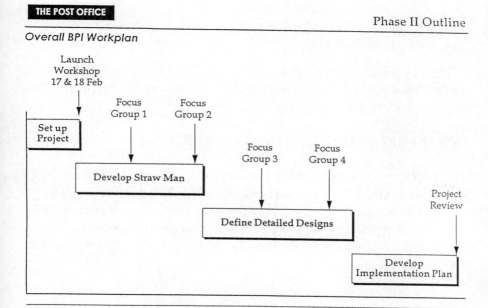

Figure 7

The aim was to include as many interested parties in the re-engineering process as possible (minimum 70), to ensure extensive buy-in and ownership of the plans.

Deliverables for the BPR project included:

- List of business processes
- Detailed process designs to include:
    - Key policies and procedures
    - IT requirements
    - Critical success factor action plan
    - Process flow chart
    - Process task definition
    - Impact analysis
    - Job descriptions
    - Organisation requirements
    - Responsibilities and accountabilities
    - Decision making matrix
    - Key perfomance indicators
- Organisation recommendations
- Implementation plan

The key principles were that one process should suffice for all product categories, despite the wide range of items purchased and supplied. Also, anything was possible!

BPR OR BPI - WHAT WAS REALLY ACHIEVED?

The BPR project started with a two-day session with a core team of 15 people to develop the strawman process. The first day was used to get people up to speed on the Phase I work and also covered teaching sessions on teambuilding, paradigms and change management. On the first evening, the key process steps for the order fulfilment process were identified.

- Customer needs identified
- Order
- Supply
- Receipt by customer
- Payment to supplier

The second day was used to break down each of the process steps using the IDEF technique. An example of one of the steps is shown in Figure 8.

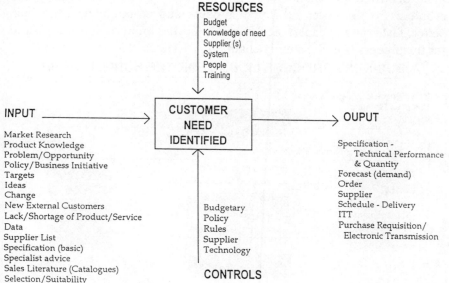

RESOURCES

Budget
Knowledge of need
Supplier (s)
System
People
Training

INPUT ⟶ | CUSTOMER NEED IDENTIFIED | ⟶ OUPUT

Market Research
Product Knowledge
Problem/Opportunity
Policy/Business Initiative
Targets
Ideas
Change
New External Customers
Lack/Shortage of Product/Service
Data
Supplier List
Specification (basic)
Specialist advice
Sales Literature (Catalogues)
Selection/Suitability

Budgetary
Policy
Rules
Supplier
Technology

CONTROLS

Specification -
  Technical Performance
  & Quantity
Forecast (demand)
Order
Supplier
Schedule - Delivery
ITT
Purchase Requisition/
  Electronic Transmission

Figure 8

Clearly at this level there is not likely to be much that requires re-engineering, so further workshops were arranged to look at process levels 2 and 3. These workshops included business representatives. Due to the tight timescales it was necessary to run workshops weekly. These were supposed to alternate between the core team and business representatives, but often people just attended when they were available. We quickly found ourselves covering the same ground at each meeting and not making progress at all, partly due to inadequate scoping and current paradigms. Suggestions for alternative ways of doing things (often from myself) were met with the response that there must be a good reason for doing things the current way. The workshop attendees also queried whether we were able or had the authority to tackle finance processes. To add to our troubles, we finally received advise about any assumptions we should be making and this effectively curtailed our radical thinking as we became caught up with historical systems, financial conventions and the budgetary and authorisation processes to be used. We had by this time identified the finance processes as those which were causing most problems to the order fulfilment process and stopping us from being better, faster and cheaper. See Figure 9 for the interrelationship diagram for the processes in the supply chain.

BUSINESS PROCESS MODEL FOR THE SUPPLY CHAIN

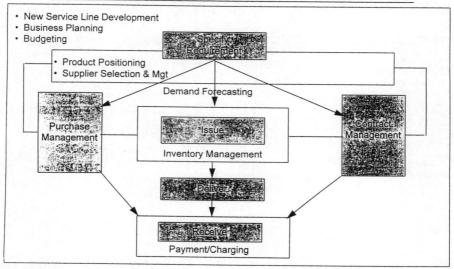

Figure 9

The new processes were finally developed to fit the current systems, but the final report which contained an overview/summary was used to detail the suggested changes we wanted to the foundation processes which were causing the problems.

The Supply Chain Review reported in July 1994 but had a number of problems with the financial business case. It then hit a lengthy period of delay as the Post Office Businesses disputed the projected savings and recommendations to close all the regional warehouses and concentrate the work at Purchasing + Logistics Services. The Product Group Teams, First Class Supplier Programme and other discrete projects continued to move forward but the Supply Chain Review stalled until early 1996 when the Businesses finally signed up the project and the benefits case. The Review is now in the Implementation Phase III.

THE LESSONS LEARNED

The first attempt at BPR was clearly not as successful as we would have hoped; although many of the ambitious recommendations are now being implemented, there was a considerable time delay. Personal research for a PhD identified some of the key problems we encountered. These problems were not peculiar to the Post Office but encountered by many other organisations:

- Senior management commitment - probably the worst problem we encountered, which led to lack of ownership of the whole Supply Chain Review by the Businesses and the lengthy delay in implementation.
- Clear vision - when we began the re-engineering exercise, we knew that we wanted the process to be better, faster and cheaper. What we did not do was to articulate how much better, faster and cheaper it should be. This allowed people to get caught into the current state paradigm because no stretching goal was proposed which would force them to come up with innovative alternative ways of working.
- Heritage systems - these are often a key inhibitor to BPR because organisations are not prepared to scrap large investments in systems where processes are redesigned and require different systems. Many organisations are now adopting what has become known as 'pragmatic re-engineering' where they will implement a new system which does 80% + of the job required rather than wait for a new system to be designed and built. However, this does not help those organisations where there is already a system in place which cannot be replaced.
- Communication - probably the most common problem of all. Many practitioners now advocate that it is not possible to do too much communication.

BPR 2 PROJECT WINWIN - LEARNING FROM OUR MISTAKES

The central warehouse had not been included during the initial BPR project for two reasons - lack of time and the new Operations Director had not been appointed. By December 1994, he and his direct reports were keen to work on their business processes. They were offered two

252

alternatives:

- Map the current processes, then look for duplication and waste and do some low level process improvement
- Consider the warehouse as a whole and take potentially radical action to improve performance. The team were cautioned that this would be a major undertaking.

The team were unanimous in their choice for radical action and a two-day workshop was set up for February 1995. My personal research into BPR had allowed me to develop what I believed to be a better methodology than that used in the Supply Chain Review. This is shown in Figure 10. Essentially it suggests that it is necessary to develop the vision for re-engineering from four different areas: a blank sheet of paper; problems with the current process, understanding of world class practice and customer/supplier requirements.

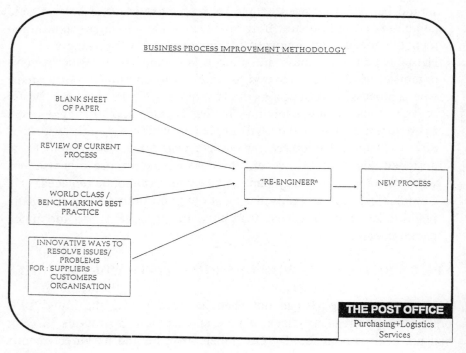

Figure 10

The two-day workshop included the top team for operations as well as suppliers, customers, and two experts on warehousing (who had already visited the site and understood the current issues and problems). We used day 1 to identify all the issues/problems and identify world class performance. From mid-afternoon, we were able to start describing a clear vision for the warehouse in the future, detailing what it would be like in terms of customer satisfaction, working environment, suppliers, employees, profitability etc. During the evening of day 1, the information was grouped so that it could be worked up into action plans by syndicate groups on day 2.

The syndicates split the list into 26 specific projects which were then broken down to identify the key activities required for success. Quickwins and longer term action were agreed and a process for moving forward. The top team split the projects between themselves, acting as leader or owner to progress the activity to completion. Each project was given a mentor (Operations or Quality & Business Process Director) to provide help/support or guidance as required. The mentor was also tasked to deal with any barriers the project owner could not resolve. Additional resource was provided by the Quality Team and a manager who had some free time. The project owners were tasked to complete as much as possible in eight weeks.

The 26 projects identified were:

- Reduce overhead costs e.g. administration support
- Move to a can-do culture
- Raise productivity
- Suppliers - range of projects, including receipt of goods
- Perfect order - 100% accuracy in order fulfilment
- Introduction of end-to-end measurement
- Re-engineer the returns procedure
- Extend opening hours
- Better, faster, cheaper - improve client relations
- Client stock improvements (for Post Office Counters)
- Obsolete stock procedures
- Reduction in inventory levels - live stock
- Better use of accommodation in the warehouse
- Employee skills matrix and training and development

- Two-way communication
- Re-engineer the complaints procedure
- Introduce service level agreements with customers
- Introduce customer satisfaction measurement card and customer care
- Improve telephone responses
- Extend pricing menu for different warehouse goods
- Re-engineer new product launches for early warehouse/purchasing involvement
- Review and recommend new management information requirements
- Develop action plan to move to paperless warehousing
- EDI paperless processing for orders and invoices
- Invoicing for third party work
- Information flexibility

The BPR project became known as WINWIN - Warehouse Improvement Now, Which Is Noticeable. The eight project owners were given no extra specific time for the project and had to fit in the extra work with their day job, yet their progress was remarkable. The two-day workshop left everyone feeling highly motivated and convinced that the changes could be made. One of the team later remarked that it was the best two-day session he had ever attended.

By the review session in early April 1995, 10 projects were either completed, implemented or due to be implemented, e.g.

- Obsolete stock procedures - all obsolete stock identified and agreement obtained from the Businesses for disposal, resulting in 10% reduction in inventory and 4% space released by volume (600 pallet positions).
- Improve telephone responses - guidelines and targets introduced for telephone answering within 5 rings), standard response/greeting. Monthly monitoring with report to ensure that all customer numbers are answered to target.
- Re-engineer returns procedure - process completely changed, so customer has replacement or credit note sent out within 24 hours (improvement from 2-3 week process).

10 projects had substantial output by July 1995:

- Extended opening hours - 4½ day week extended to 5 day week in June 1995. Proposals under discussion for move to annual hours and increased flexibility to cover customer requirements.
- Employee skills matrix and training and development - introduction of an appraisal process for front-line staff identifying all core relevant skills, both technical and social. Skills matrix developed to show current skills and those areas where training is required.
- Introduce customer satisfaction measurement card and customer care - questionnaire postcard developed with customers; now included in all orders despatched from the warehouse. Report shows 10-20% improvements in satisfaction with uniform orders within the first two months. Specific new posts created in uniform and print teams to act as customer care/liaison. Customer response indicates that this has been very well received.

6 projects had made substantial inroads towards their desired state by September 1995 e.g.

- New product launches - both Royal Mail and Post Office Counters business processes changed to ensure earlier involvement from purchasing and warehousing to allow input to the specifications for the product/support material. Potentially huge savings on cycle time reduction and spend. Awaiting first live trials.
- Use of accommodation - identified saving of 30+% by moving all the current work into two warehousing units instead of three.

In July 1996, there was a major review of WINWIN, leading to a sign-off of the original project and absorption of outstanding work into day-to-day business. The review session (with some new players, due to organisational changes) led to a revitalisation of WINWIN and the launch of another 26 projects.

The last three years have seen dramatic changes in the Post Office supply chain, moving from an internally focused, bureaucratic department to a customer facing service with fledgling world class, leading edge processes. The move to managing the supply chain as a process instead of a set of discrete functions is not complete and there is a lot more work ahead, but we remain convinced of the benefits of the changes we are making.

Business process re-engineering is not something that any organisation should take on without understanding the work involved and the depth of the changes required, but with an exciting vision for the future and senior management commitment, it can be done successfully.

CHAPTER 13

Implementing Concurrent Engineering

B. M. Brooks *and* S. G. Foster

1 BACKGROUND

The concepts which underpin Concurrent Engineering (CE) have been described in many papers and books. The aim of this case study is to take those concepts and describe how they have been applied to bring about beneficial change in a large engineering company. This description of the implementation of Concurrent Engineering is based on a case study of Vickers Shipbuilding and Engineering Ltd (VSEL - now part of GEC Marine) which is the largest of three naval shipbuilding yards in the UK.

VSEL is the only shipyard capable of building the full range of vessels for their main customer, the Ministry of Defence (MOD). Their business strategy states an intention to maintain the leading position in the supply of submarines and large surface warships.

Effective exploitation of Concurrent Engineering with multi-disciplinary Design and Build teams and the use of modern CAD modelling and Engineering Data Management (EDM) techniques were identified as key essential features to support future contracts.

This case study describes in detail the work undertaken by VSEL and PA Consulting Group, to introduce Concurrent Engineering at VSEL. This has been undertaken in three distinct phases, illustrated in Figure 1.

There are many definitions of Concurrent Engineering (CE) and the term is widely used but often means different things to different people. One definition which is commonly quoted is from the US Department of Defence, Defence Analysis Report, R-338 :

258

Concurrent Engineering: A systematic approach to the integrated, concurrent design of products and their related processes, including manufacture and support. This approach is intended to cause the developers, from the outset, to consider all the elements of the product life cycle from conception through disposal, including quality, cost, schedule and user requirements.

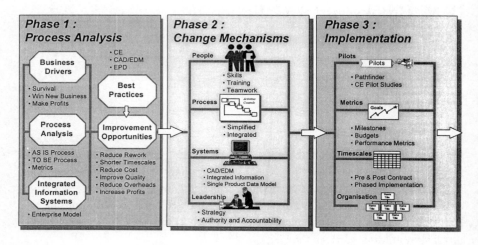

Figure 1   *The Introduction of Concurrent Engineering - the Three Phases*

This definition is fine, as far as it goes. It concentrates on the process elements of Concurrent Engineering, but makes no reference to the way in which people work together to make it happen or the tools which they must have to facilitate the process. Typically, other definitions are either very broad and general or have a specific focus on teamwork, technology and systems or a particular technique such as Total Quality Management (TQM), Quality Function Deployment (QFD) or Design for Manufacture and Assembly (DFMA).

In practice a successful approach is likely to include many or all of these ingredients. A distinguishing characteristic of CE is that it integrates the use of multiple sources of knowledge to enable superior and timely decisions to be made about all stages of the product lifecycle.

2 PHASE 1 - PROCESS ANALYSIS

2.1 Key Business Drivers - the need for change

The overriding business issue driving the need for change at VSEL was the **survival** of the Company.

A strong technical capability differentiates VSEL from other UK shipbuilders and, although significantly reduced in size over former years, the Company has maintained core technical skills and competency at a time when other shipyards have gone into terminal decline.

A core technical capability is an essential pre-requisite to retaining a licence to build nuclear submarines. Technical strength in design, test and commissioning areas, are necessary to support current contract commitments but were equally important in tendering for and winning new business. A major driver for change was thus the need to maintain and enhance the technical capability. If correctly applied this would lead to success in **winning new business,** essential to the company and the second major business driver.

A third key business driver was **increased profits.** The adoption of people, process and system changes, coupled with the appropriate leadership and commitment, were recognised as appropriate change mechanisms to yield significant benefits in the following areas :

✓ Improved Quality   ✓ Reduced Direct Cost
✓ Shorter Timescales   ✓ Reduction in Rework
✓ Reduced Overheads   ✓ Increased Profits

VSEL has an established reputation for high quality products. The introduction of CE was a necessary ingredient in further reducing costs and increasing efficiency, whilst continuing to produce products which satisfy or exceed customer expectations.

2.2 The Changing Business Environment

VSEL operates in a rapidly changing business environment. Survival requires vision and commitment to improved operating efficiency and effectiveness.

In recent years the major customer, the Ministry of Defence, has changed its procurement approach to major contracts, through the introduction of competition for firm price contracts and the devolution of risk, through the introduction of Prime Contractorship. VSEL responded by developing a Business Strategy which includes a major involvement in both UK Naval Submarine and Surface Ship programmes. Strategic partnerships with other companies have been adopted as a means of sharing the risk and maintaining an effective skill base.

VSEL has to demonstrate leadership in the application of technology and other key aspects of contract management necessary for successful completion of future contracts at minimum acceptable risk.

2.3 Opportunity and Timing

The opportunity to make an impact on a new major design and build contract occurs very infrequently. The last significant design cycles occurred in the early 70's and mid 80's. Major investment in the technical areas came as part of that cycle, with the introduction of CAD in the early 80's. At that time the technology lacked appropriate breadth and depth of functionality. Not surprisingly the benefits from the investment were not fully realised and to some extent the technology got a bad name. The use of CAD was relegated, in the main, to that of an electronic drawing board. The quality of the design output led to rework in production against drawings which were often late and failed to adequately reflect production requirements.

The capabilities of today's CAD systems are beyond the wildest dreams of the early CAD practitioners in terms of 3D modelling and visualisation capabilities. Links to other production systems such MRP and Project Planning are now readily available.

In recent times, the opportunity for VSEL to tender for major new contracts has provided the catalyst to re-engineer the processes and apply new CAD and associated EDM technology, to gain business advantage through a change in working practices.

Strong leadership, good communications and commitment were required to ensure these changes were achieved and became self sustaining with no looking back or reverting to old ways as part of short-term fixes.

For VSEL, the combination of a changing business environment, increased competition, new contract opportunities and the need to retain a strong technical capability, all coincided to create the need for major changes to the efficiency and effectiveness of the business.

2.4 Business Improvement Programme (BIP) - the response

In responding to the identified need for improvements in efficiency and effectiveness, a Business Improvement Programme was established. An initial study defined the structure and scope for the BIP initiative.

The aim of the Business Improvement Programme was to facilitate the development of : *"A World Class Company achieving its corporate objectives through **empowered people**, using **simplified processes** and supported by **integrated information systems**"*.

The BIP initiative encompassed four cross-company projects:

- Winning New Business
- Program Management
- Effective Management of Design
- Managing Material

Figure 2 indicates how the four cross-company projects integrate and identifies some of the other enabling projects.

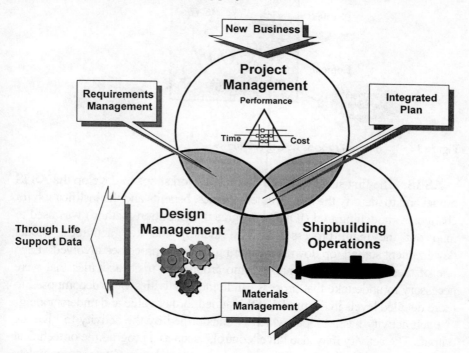

Figure 2   *Elements of the Business Improvement Programme*

The remainder of the case study describes the Implementation of Concurrent Engineering as part of the Effective Management of Design element of BIP.

2.5 Business Process Re-engineering

Business Process Re-engineering techniques were used to identify changes. The aim was to characterise the current way of working (or the **'AS IS'**), to develop a model of the future way of working, (the **'TO BE'**) and to progressively implement the practical use of the new tools, techniques and improved ways of working through **pilot projects** using real work tasks. The implementation process is illustrated in Figure 3.

PROCESS RE-ENGINEERING

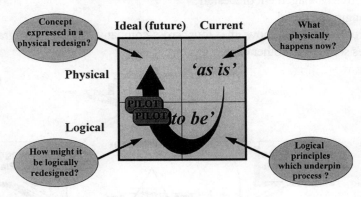

Figure 3   *Business Process Re-engineering*

AS IS: The first stage involved a series of workshops to develop the AS IS model and to identify the improvement themes, benefits and the justification for change. A modelling tool (IDEF - Integrated DEFinition method) was used to map out the main processes and to identify the key activities. Model development started top down, considering the main processes involved. Each one of these was then decomposed into around six main activities that were necessary to undertake that process. Individual activities were decomposed to more detailed levels in the model, if this added further value and understanding.

Each activity has an input which is transformed by the activity to give the output. The activity may also have 'controls' such as a programme or technical standards and an indication of the resources or 'mechanism' used to complete the tasks. Figure 4 illustrates the form and structure of the IDEF models.

The development of the 'AS IS' model took place through individual and group model building sessions where the main processes and their associated activities or sub processes were brainstormed and grouped together. Considerable use was made of 'Post it' notes to assist in moving activities and tasks around and in developing the appropriate groupings. Models were developed by working both top down and bottom up to reconcile any differences. As these models of the process converged, a common view was progressively developed using workshop sessions to confirm that the model was representative of the 'AS IS' situation.

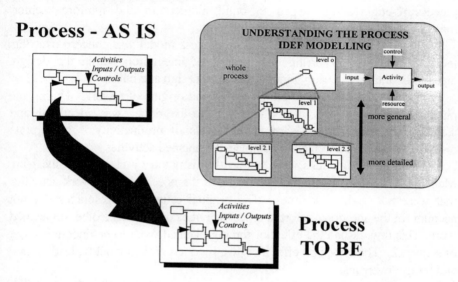

Figure 4   *Process Modelling - IDEF*

In creating the model, both the needs and opportunities for change began to emerge. An initial view of the way forward was formed and cross functional teams worked together to identify any sub optimal or non value adding activities. To make effective use of process models to stimulate change, team involvement in the development and ownership of the AS IS model was vital. The dissatisfaction or frustration associated with the current process provided the momentum for change.

The IDEF models represented a very effective way of describing processes and establishing any gaps, issues, bottlenecks or opportunities for improvement.

TO BE: The next stage was to develop the 'TO BE' process model. This used benchmarking data to ensure the re-engineered processes reflected 'industry best practice'. A series of improvement themes and metrics defined best practice Concurrent Engineering together with the essential changes in the performance of the people, process and systems. All of these were inputs to the 'TO BE' development.

The process models and the associated measures (value added, rework, gaps, issues, bottlenecks and opportunities etc.) help identify improvements for the future.

Implementing 'quick hits' as a means of demonstrating early benefits from process re-engineering, helped to build momentum and reinforce senior management support for the change initiatives.

An issue identified at all levels of the process model was concern over the levels of 'rework' occurring due to lack of integration within the design disciplines and between key functions such as design and build.

Part of the analysis identified the extent and origins of rework. During the planning of project activities, the work 'to be done' was identified and assumptions were made regarding the levels of productivity. Appropriate resources were then allocated to complete the planned activities.

These planned activities were frequently disrupted by changes in the plan. Many of the changes were due to rework or the need to repeat work activities that were not 'right first time'. The original estimate of resources did not account for the additional effort to repeat activities or undertake other unplanned work. This rework had the effect of forcing planned work to be undertaken out of sequence. The knock on effects of this, if not controlled quickly, lead to time and budget overruns.

The discovery of rework late in the process greatly magnifies the impact of change, with serious consequences if penalty clauses are involved. The improved TO BE process recognised the existence and impact of the rework cycle, illustrated in Figure 5.

Part of the analysis of the process identified specific improvement themes which would become the focus for applying change during the pilot projects. The improvement themes are listed below under headings related to process, people and systems. These same topic headings have also been used as part of a CE behavioural audit. This assessed the level of recognition of the TO BE way of working and provided a base line for measuring improvements in the adoption of CE.

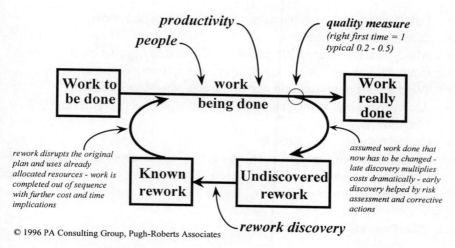

Figure 5 *The Rework Cycle*

The improvement themes are:

Design Processes and Planning
- Requirement / Acceptance Management
- Product Development Strategy (including integrated strategies for design, build, procure and support)
- Planning, structured process and 'toll gates' to control rework
- Looking ahead / accommodating likely changes / impact management
- Integration (performance, space, weight, unit cost)
- Early consideration of production engineering and support
- Change Management

Teams and Organisation
- Multi-disciplinary teams, focused on 'process'
- Clarity of roles and responsibilities
- Partners and suppliers part of process as team members
- Active, motivated and co-located teams (planning to avoid rework)
- Project / programme management linked to critical resource management
- Training and development in critical skills

Information and Systems
- Managed requirements, documentation control and compliance

- Data navigation, accessibility, accuracy, timeliness and status
- Data format and interfaces
- Ability to accept and update partial / changing data / configuration control
- Parametrics to facilitate design iteration with minimum re-creation
- Tools for integration and simulation - space, weight, unit costs etc.
- Ability to manage structured partial release of product definition and Bill Of Materials (BOM)
- Critical resource management
- Management of support data, (CALS - Continuous Acquisition and Lifecycle Support and a CITIS style environment - Contractor Integrated Technical Information Service)

2.6 Pilot Projects

The TO BE model was initially a theoretical representation of how the company could operate in the future. Benefit projections defined the metrics, savings and necessary investment. The TO BE model showed the process was different but was it better ? People frequently have to experience a new way of working before they will believe the level of improvement possible.

To test the proposed changes and translate the theory into practical experience, a series of 'pilot projects' were defined, planned and launched. These projects focused on the 'pressure points' of the TO BE model. They involved project related work undertaken typically over a three month period, with clear objectives, budgets, plans and deliverables. The pilots were used to test a combination of the people, process and systems changes and to confirm the anticipated levels of benefit.

Part of the preparation for the pilots involved selecting and mobilising the teams and training them in the new way of working together, with relevant techniques such as process modelling, milestone planning, teamworking and problem solving.

A prime objective of the pilot projects was to start to do things differently. Positive results and evidence from pilots reinforced the confidence of the senior management group. Their support for the change process was illustrated through their endorsement of further pilot projects.

Pilot projects provide an opportunity to challenge the status quo. They also have to pass through a difficult phase with some people remaining sceptical and resisting change since their experience is being challenged. A key trap to avoid is to end up spending time and effort arguing and trying to gain acceptance for complex, theoretical future models. The broad direction of the 'TO BE' has to be justified and the potential benefits demonstrated. After that, all efforts should focus on practical experience of bringing about and sustaining change.

Typically, as more people participate in pilots and are trained in the new way of working, the existing organisation comes under strain. It will need to be 'reshaped' to accommodate a project and process focus, with multi-disciplinary teams operating within a project and functional resource matrix organisational structure. As more pilots are undertaken the form of the re-designed processes takes shape and progressively, the new way supersedes the old.

It has been shown that multi-disciplinary teams can save 30% on costs and up to 50% on overall lead time. At the early stage in the transition, such figures are open to criticism from sceptics, such as *"this improvement was due to it being a small project with senior management attention"* - a form of 'Hawthorne Effect'.

Evidence shows that a multi-disciplinary team, motivated to achieve improvements, can achieve significantly better performance.

Process changes are, however, only one of the key ingredients in delivering business improvement. If these teams are then supported by a very effective CAD modelling system, access to accurate and timely information via an Engineering Data Management system and a well planned and implemented Concurrent Engineering approach, then the benefits can be substantial. The aim of pilot projects is to add substance to this 'act of faith' by the involvement of teams working in a new way, utilising new tools and techniques. Challenging time and cost targets were set to force people to *do things differently.*

The challenge for the management team was to share the lessons learnt in the pilots and to create the right conditions for the new approach to take root and flourish. The line management role changes from controlling the work to 'coaching' the multi-disciplinary teams, as part of a 'learn and share' approach to change implementation.

2.7 Communication

Throughout the whole process, leading to the development of the TO BE model and the implementation of the changes, the importance of frequent and effective communications was recognised.

Considerable effort was applied to ensure that other staff and teams not directly participating in the modelling process had the opportunity to review the latest thoughts and contribute comments from their own experience. 'Looking back' sessions were held with various customers of design information to seek their ideas on the changes in the organisation and operation of the technical processes that would assist them in becoming more effective and efficient.

The change teams also worked with a senior management steering group, to ensure all members were fully up to date with progress and supported the proposed changes. The need for good communications was reinforced by active participation. Individual managers were required to underwrite the proposed levels and sources of benefit.

In this section it has only been possible to introduce some of the techniques of process re-engineering employed at VSEL. The particular route adopted by an individual company will vary and may be driven by a particular theme such as time-to-market, achieving best practice or a variety of cost and quality related themes.

For VSEL, the main focus was to enhance and secure the core technical skills by the introduction of a comprehensive Concurrent Engineering approach and to improve efficiency by reducing rework.

Whichever theme drives the need for change, the ingredients of the approach described are relevant and provide a framework for modelling, assessing, re-designing and implementing improved processes. This framework is summarised below.

Objectives and focus	agree the need and areas for change
'As is' modelling	understand the current process & performance (IDEF model)
Process logic	understand the logic that underpins the process
Process re-design	- training in tools and techniques including Concurrent Engineering - benchmarking to set improvement targets - 'to be' Modelling describing the direction and vision
Set up change teams	sponsors, potential pilots, change mechanisms, training
Communication	senior management and user commitment
Pilot project	start experiencing doing things in a new way
More pilot projects	roll out the 'new way' for processes, systems & organisation
Continuous improvement	coach, support, measure and monitor progress

3 PHASE 2 - CHANGE MECHANISMS (see figure 1)

Change mechanisms are the enablers or 'things you have do' if the organisation is to change and progressively adopt a Concurrent Engineering approach. For example the 'pull' of information by the downstream team members has required the application of people and process change mechanisms.

In a similar way, the use of effective EDM and CAD tools will enable project team members to better communicate and share information. Technology can improve control and visibility of the evolving design and illustrate how the interrelationships between the many different contributions to the complete product definition can be managed. At VSEL, a Technology Demonstrator has been combined with people change mechanisms to support performance improvement themes associated with modelling, data management and controlled parallel working. This provided a 'greenhouse environment' for evaluating changes.

The Change Mechanisms used to promote a new way of working, were developed in Phase 2. As more of these people, process and system change mechanisms are applied through pilot projects, the Concurrent Engineering way

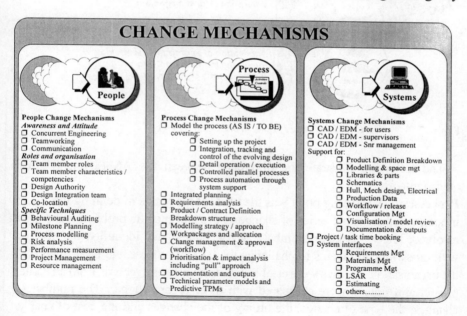

Figure 6   *The Change Mechanisms*

of working develops.

A selection of the people, process and systems change mechanisms used and planned is illustrated in Figure 6.

3.1 Achieving Concurrent Engineering

It is the combination of people, process and system changes, that define the 'things you have to do' if the organisation is to create a Concurrent Engineering environment. This combination is illustrated in Figure 7.

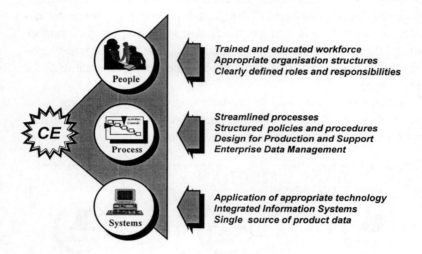

Figure 7   *The Combination of People, Process and Systems*

The following illustrate the difference between the AS IS and TO BE environments:

Processes: The starting point was the work of the teams in defining the AS IS model, together with a series of change issues and specific problem areas to be resolved, known as 'sore thumbs'. These issues were identified by working with cross function teams to challenge the current process and performance of the organisation. Metrics were established using data and records from an earlier project and figures were compared with other organisations. The number of changes, the type of change, the timing of the changes and the cost of change were used to build a model indicating the levels of rework.

The TO BE model incorporated new processes to resolve or eliminate the 'sore thumbs' and reduce the levels of rework that constrained the performance of the organisation.

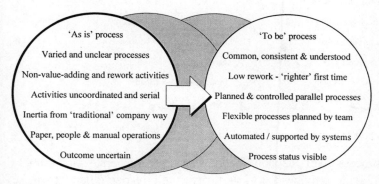

Figure 8 *Process Changes*

These revised processes concentrated on improving the integration between disciplines and the level and control of concurrent working. The aim was to manage the convergence of an evolving design within set performance, cost and time parameters. Some of the anticipated process changes, in moving towards a Concurrent Engineering environment, are illustrated in Figure 8.

A key characteristic of an improved process is the 'pull' of required information by the downstream team members rather than the traditional 'push' of documentation from the upstream functions. This allows some activities to start earlier and others, where appropriate, to be undertaken in parallel.

People: Process changes are implemented through people working together in a different way as illustrated by the changes in Figure 9. Workshop sessions were used to test and communicate an understanding of the TO BE process. They also discussed how behaviours would need to change and how people can work together more effectively in teams that focus on using the simplified processes. When building teams, consideration of individual skill profiles and the overall mix of capabilities is important. The concept of 'T' shaped people seeks to match individuals with an in depth specialisation (the vertical part of the 'T'), with others that provide an overlapping breadth of more general skills (the horizontal part of the 'T').

Many aspects of design, production and support people working together require the frequent interchange of information. Co-located teams using a

272

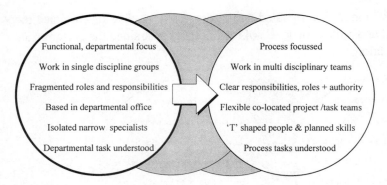

Figure 9   *People Changes*

streamlined process will depend upon access to accurate and timely information. If some activities are to start earlier and overlap others, it will be essential to give all team members access to the 'current assumptions' and design status.

An early and very significant part of the work at VSEL was devoted to defining the supporting systems environment and application tools that would be essential to underpin the changes in the processes and people. In particular, this environment had to support concurrent working and people sharing information.

Systems: Increased up front effort and working concurrently with partial information involves more iterative changes. Making these changes in a controlled way that is visible to all the team was identified as a crucial ingredient in executing contracts successfully, in less time with reduced rework. The technology and information management aspects of a CE environment were addressed by the systems dimension and the changes are illustrated in Figure 10.

A number of databases existed to manage records. A key element of the TO BE process was to move from basic records management to a systematic way of structuring and navigating through data. This facility has to be supported by a workflow process that helps to allocate priorities, enables a 'pull approach', manages the progress status and controls access.

The use of effective Engineering Data Management (EDM) and CAD tools are the key enablers that permit project team members to better communicate and share information. This gives improved control and visibility of the evolving design, and the ability to track the interrelationships between the many

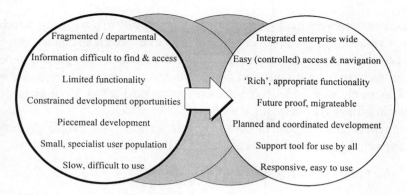

Figure 10   *System Changes*

different contributions to the complete product definition.

The AS IS model confirmed that many processes relied upon manual paper based 'systems' to document progress. The tasks associated with maintaining this information consumed high levels of valuable time from key resources. Improved processes underpinned by effective systems can eliminate, automate or streamline some of these tasks and help to ensure that design changes and the appropriate versions of data are available to all intended users of the information at the right time.

A system-based technology demonstrator was used to test out some of the principles of EDM including the product structure approach and workflow processes. This demonstrator proved invaluable in helping to communicate the approach to customers, senior management, design, production and support groups.

It also provided the foundation for the development of a Statement of User Requirements (much more interactive and effective than circulating long, complicated written descriptions of a theoretical system). It has helped to quantify the levels of effort necessary to configure a system for VSEL and has demonstrated the potential benefits which have subsequently been used to support the business case for investment in CE and CAD / EDM systems.

The combination of people, process and systems: Any of the people, process or systems changes described above would bring about some improvements and initial benefit calculations have considered each in isolation. It is however, the combination of interrelated improvements that creates a true Concurrent Engineering environment and the leverage of the business benefits.

274

Revised processes are made possible by *people working in a new way,* which requires the *sharing of information* and effective communications using appropriate technology and systems. To make effective use of the people, some processes and communications can be *automated* or supported by systems rather than using paper and non value-adding manual effort to re-input or transcribe information from multiple sources.

Concurrent Engineering can be depicted by the integration of people, process and systems change mechanisms as illustrated in Figure 11.

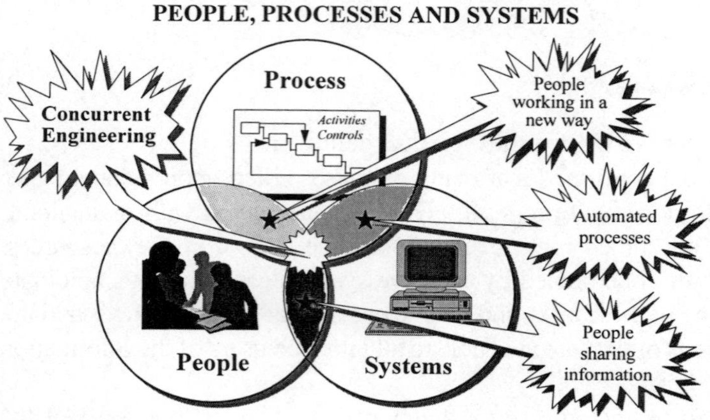

PEOPLE, PROCESSES AND SYSTEMS

Figure 11   *Concurrent Engineering*

As the people, process and systems changes become embedded in the 'normal way of working', the central Concurrent Engineering 'star' in the figure expands to encompass the individual elements. In an organisation where Concurrent Engineering has 'matured', the desired behaviours, processes and use of information are all part of the natural way of executing contracts or developing products. All participants do the 'right things' without question and expect their colleagues to do the same.

In most companies these changes take 3-4 years to reach an acceptable level of maturity. It doesn't end there however, for change is a continuous management process and the involvement, support and commitment of all the management team must be maintained throughout. This starts with support for awareness training and is practically tested and demonstrated as pilot projects start the implementation of change.

Any company facing world wide competition and wishing to improve their performance should pay serious attention to the people, process and systems changes implicit in adopting a Concurrent Engineering approach. If properly applied, CE is capable of yielding significant business benefits of the following order:

$\Rightarrow$ reduced time from concept to delivery of a product (often 30% - 50% reductions)

$\Rightarrow$ improved quality, responsiveness and conformance to customer requirements

$\Rightarrow$ reduced costs of product development, delivery and support (often 15% - 30%)

$\Rightarrow$ reduced rework and 'non value-adding' activities

$\Rightarrow$ increased profits

Some of the key features of a best practice Concurrent Engineering environment involve the contributions made by people, processes and systems:

Customer Driven Activities

✓ customer involved in process
✓ clear definition of requirements
✓ good understanding of the market

Lean Processes

✓ integrated strategies and plans
✓ focus on integrating activities
✓ information 'pull' from downstream rather than 'push' from upstream
✓ controlled and measured parallel working

Motivated Teams

✓ multi-disciplinary teams
✓ co-location
✓ focused roles
✓ single point responsibilities
✓ authority and empowerment
✓ excellent skill base
✓ recognition and rewards

Timely and Accessible Information

✓ best practice tools e.g. CAD & Engineering Data Management
✓ visibility of information and documentation
✓ visible progress status
✓ selective access to product data
✓ reuse of existing product data
✓ single master source of product data
✓ Configuration Mgt & Change Control

3.2 Systems Support for Concurrent Engineering

Two main system related tools were identified as being essential to support a Concurrent Engineering environment and provide the timely and accessible information. These were:

An **Electronic Product Definition (EPD)** system, created through the application of Computer Aided Design (CAD) techniques and processes.

An **Engineering Data Management** system, which electronically holds a single source of Product Definition in the form of CAD models, drawings, reports, specifications, part attribute data etc. and makes this structured information immediately available to those who need it.

3.2.1 Electronic Product Definition (EPD)

The Electronic Product Definition is an assembly of all the data needed to fully define the logical and physical representations of the product. The data is held in electronic format and is created and viewed through CAD and associated systems. A product could be anything from a detailed engineering component such as a valve, to a complete product assembly (in VSEL's case this could be a complete submarine or ship).

The role of a comprehensive full scale 3D CAD model in the integration activity associated with the development of complex products was identified as being crucial to the effective management of the design process.

One of the prime objectives in applying 'best practice' CAD technology to all future contracts, was the eventual elimination of expensive plastic models and the delays involved in updating and freezing the model during review periods. The 3D electronic model and visualisation techniques allow reviews of the design layout, ergonomic evaluations and support reviews to be undertaken long before the actual product takes shape or any significant production investment has been made.

The ability to share model data interactively supports the adoption of the desired Concurrent Engineering environment. Multi-disciplinary teams develop the design with full visibility of its evolution and with an effective means of communication between team members. An example of this environment is illustrated in Figure 12.

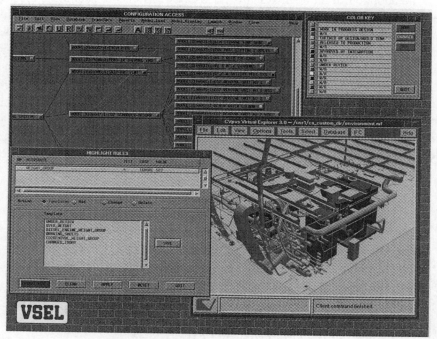

Figure 12   *Electronic Product Definition Environment*

Application of EPD technology enables rapid design development and provides flexibility in the interrogation and manipulation of the geometry. The Manufacturing Bill of Materials can be created in time to support effective modular construction and assembly processes.   Direct data links with numerically controlled machine tools facilitate electronic transfer of geometric and toolpath data in support of fabrication and manufacturing activities.

Digital pre-assembly techniques are used to prove the integrity of designs electronically, prior to the expensive commitment to production, thereby reducing rework, lead-times and minimising costs.

By creating and maintaining a single source of product data, the timely release of accurate data to 'downstream' processes is possible, further reducing rework and improving production productivity.   Effective engineering change control and configuration management are achieved through controlled access to the single Electronic Product Definition.

The adoption of an open systems philosophy has enable data to be created and shared, both internally within VSEL and externally with major sub contractors and the customer. The use of CAD combined with multi-media communication devices and video-conferencing is being explored to encourage team working and reduce constraints imposed by geographical locations. Compliance with CALS (Continuous Acquisition and Lifecycle Support) initiatives and STEP (STandard for the Exchange of Product model data) standards will ensure that data created in design can be used to support 'downstream' process and ultimately form part of the Logistic Support Analysis Record (LSAR) and technical documentation of the product.

The benefits of an Electronic Product Definition are seen to be:

✓ Reductions in time to market
✓ Reduced cost of change, error and rework
✓ Lower development costs
✓ Increased Product quality - 'righter' first time
✓ Wider and faster access to design information
✓ Reduced manufacturing documentation
✓ Data re-use and data sharing

3.2.2 Engineering Data Management (EDM)

The EDM environment manages all of the data and associated information relating to the evolving product definition. It allows exchange of product related data between engineering, planning and manufacturing software applications. The same EDM environment provides the capability to track the customer requirements and design status throughout the development processes, up to the point of acceptance and beyond, into service.

An integrated and open environment allows a user to access any information required to undertake their duties from an engineering workstation or PC. Computer applications support the analysis, review, approval and distribution of product information (workflow), and sophisticated search facilities provide instant access to related reference and guidance information.

A 'design chronicle', maintained by the system, ensures that the design intent, methodology and history are captured. Electronic messaging informs interested parties that a particular event has occurred and that follow-on tasks should be initiated as part of the controlled workflow process. Notifications, via electronic mail, ensure that everyone who needs to be aware of what is happening, receives the information in a timely fashion.

Project and Team Managers are provided with tools to create workpackages containing both the task and all the required data. These are initiated by work instructions, progress is monitored and the product configuration is managed through the product structure. Facilities are provided to enable the technical contents of the product to be reviewed and approved on-line. The release status of the product is managed within the EDM environment and reporting tools enable monitoring and control of the product development.

All information is stored electronically in a virtual single database and retrieval of information is achieved by reference to a Product Structure, which also provides the means of 'navigating' around data. The appropriate computer applications are launched automatically when an item is selected from the Product Structure. Assigned task data is selected from the information in the workpackage.

A standard Graphical User Interface (GUI) provides a consistent method for users to interact with applications in the EDM environment and graphical tools display the relationship between the product and its constituent parts.

Subject to the appropriate authorisation, users are able to access information created on any application, from any workstation resident on the network that is capable of running the required application.

The benefits of an Engineering Data Management system are :
✓ Single source of product data
✓ Timely and accurate release of information
✓ Full traceability of design development process
✓ Effective process monitoring and control
✓ Automated system administration and data access controls
✓ Graphical data search and navigation capability
✓ Ease of use through standard graphical user interfaces
✓ Facilitates enterprise data management
✓ Supports the creation of a 'virtual teaming' environment
✓ Enables re-use of corporate data

4 PHASE 3 - IMPLEMENTATION (see figure 1)

The definition of the TO BE model and the identification of the main improvement themes completed the work under Phase 1, covering the Process

Analysis. Phase 2 focused on the 'Change Mechanisms' based on the People, Process and Systems approach and the development of training materials, teamwork and Leadership. Phase 3 addresses the progressive implementation through a series of pilot projects.

4.1 Pilot Projects

To improve understanding and confidence, achieve user and management commitment, and reduce risk, carefully selected pilot projects are being undertaken in support of 'real work' on VSEL's main contracts and bids.

To implement the TO BE way of working across the business, the pilot projects have been planned to incorporate a combination of Change Mechanisms and Improvement Themes to deploy the new way of working.

Pilot projects have been used in the following way - the pilot project team apply the change mechanisms to some specific task deliverable on the project. The tasks have a performance improvement objective that is linked directly to benefits. The management team must sign up to the savings and drive the pilot project to realise the changes and improvements in performance.

The Change Mechanisms, when applied to improvement opportunities, can generate benefits, for example:

A multi-disciplinary team *(people change mechanism)* focused on the improved process *(process change mechanism)*, using an effective CAD and EDM system for modelling space allocation *(system change mechanism)*, will ensure better integration of the design, build and support activities *(improvement opportunity)* and reduce the rework costs *(a benefit area)*.

Figure 13 illustrates the linkage between the management team sponsors, the benefit area contribution for a project and the bringing together of a number of Change Mechanisms.

These pilots help to introduce the new way of working, generate benefits and will enable the benefit estimates to be refined. A key role for the management team will be to 'own and support' these pilots and to transfer the lessons learnt. They must also propagate the changes in people, processes and systems across other projects and the organisation.

The selection of the pilot work activity, the teams and the relevant Change Mechanisms was thus critically important, especially for the early projects where 'resistance to change' or 'reversion to type' can be a threat to their success.

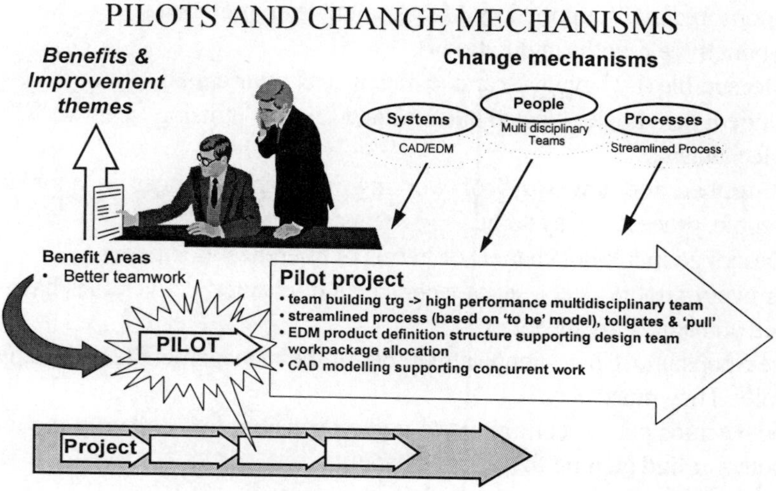

PILOTS AND CHANGE MECHANISMS

Figure 13   *Pilot Projects - Change Mechanisms Applied to Improvement Themes*

Performance measurements are a key feature of each of pilot project to ensure that an assessment of benefit can be made and measures of performance for individual design tasks can be established. Each project is controlled through a milestone-based plan against which a budget has been allocated. Dedicated resources have been allocated to each pilot to ensure its success and achieve the development of the desired multi-discipline project-based organisation.

The change team have established the following stages in the setting up and running of a pilot project:
- selecting the pilot
- mobilising the pilot team
- facilitating the pilot
- auditing the progress and outcome

Each has a checklist and set of guidelines. For example the definition of a work activity suitable for a pilot must fulfil the following criteria:

Pilots projects are:
- Clearly defined areas of 'real' project work
- Packages of work that Management team members 'need'
- Linked to a benefit area

- Sponsored and supported by Management team member(s)
- About three months in duration
- Measurable (by having clear and meaningful metrics)
- Undertaken by a selected and trained 'team' utilising specific 'Change Mechanisms'
- A testbed for new ways of working involving changes in approach to people, process and systems
- Described in a 'pilot charter' or 'terms of reference' document

For every pilot project a set of technical and 'change' deliverables have been defined and agreed by the team.

The Sponsor or Champion is a member of the management team and plays a key role. They must:

- be a senior person committed to and responsible for achieving the change outcome and gaining the targeted benefits
- support the team's desire to work in a new way
- ensure the team receives support, training and on-going facilitation
- be prepared to defend or protect the team and their 'radical approach' in the face of pressures from the rest of the conventional organisation
- communicate and promote the new way of working

The Team Leader for a pilot is responsible for the technical deliverables of the project and has been trained with the team in the appropriate change mechanisms and the Concurrent Engineering principles.

In all cases the Change Mechanisms will be deployed in the context of a pilot project: ie a real package of work in which the 'to be' way of working and appropriate Concurrent Engineering principles are applied to produce a measurable project deliverable.

Benefits: The benefit areas identified are listed below. These must be developed into a complete benefit model with the specified savings owned by individual members of the management team.

1. Reduced Project Timescales
2. Improved Departmental Efficiency
3. Greater Resource Effectiveness
4. Less Rework ('righter first time')
5. Reduced Risks and improved Confidence
6. Reduced Material Usage
7. Lower Material price
8. Improved Commercial Conditions (full recharge / reduced warranty claims)

9. Volume and Throughput

10. Reduced overheads

A 'rolling wave' of pilots is anticipated as more of the change elements become progressively embedded within the 'normal way of working'.

4.2 Implementation through Pilot Projects - The 'Pathfinder' Project

The Pathfinder project (Figure 14) was one of the first of the pilots and serves to illustrate the approach adopted to the launch and completion of any pilot. Pathfinder was a risk reduction exercise specifically designed to test key assumptions about aspects of the CAD / EDM technology, which were vital to the future success of the changes which VSEL is undertaking. Whilst focusing on the systems changes, Pathfinder also addressed other people and process issues.

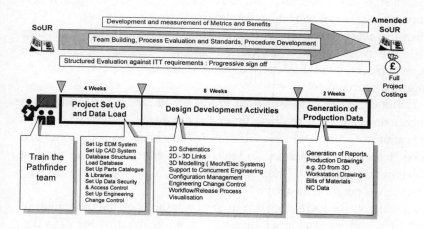

Figure 14   *The Pathfinder Programme*

It was a joint exercise with the chosen CAD/EDM vendor to demonstrate the functionality offered by the latest technology and to ensure that this would satisfy the requirements of VSEL for future submarine and surface ship design and build contracts. During the pilot project, VSEL representatives undertook 'real and representative work' on two major projects. One was at a concept stage with an evolving design and the other, an established design undergoing change. In parallel, an evaluation exercise was undertaken to assess the selected system and the quality of vendor support, against a detailed Statement of User Requirements. Performance metrics and benefits were also recorded against specific tasks and deliverables.

Figure 14 illustrates the scope, content and timing of the Pathfinder. A comprehensive ten day training programme was undertaken by the core team of twenty people. They formed multi-disciplinary design and build teams addressing surface ship and submarine design aspects.

The training combined the introduction of tools and techniques through classroom and syndicate group sessions, with the application of the technique (such as IDEF process modelling, Milestone Planning or Risk Assessment) to a specific Pathfinder task. This tested the understanding of the technique and started to get the individuals working as a team to apply the technique in a useful and practical way. A further 30 people were trained, and performed specialist support roles, assisting the core teams with the evaluation and completion of in excess of one hundred tasks.

The first four weeks were concerned with the setting up, procedures and data loading. The next period exercised the various features, processes and application tools and the final period focused on the outputs for design, production and procurement.

The 'Pathfinder' evaluation was used to test the proposed methodology of organising, managing and controlling future pilot projects. Pathfinder required the application of many of the Change Mechanisms and demonstrated improvement opportunities leading to real business benefits. Being the first pilot project it also acted as a 'shop window' giving high visibility to the way VSEL intends to operate in the future.

The organisation for Pathfinder was used to test the operating principles of improved front end integration as identified in the analysis of the 'TO BE' design process and as part of the operation of a Concurrent Engineering environment. Typically, this role balances a requirement to technically progress the design and development against business and programme implementation requirements.

The equivalent task performed in Pathfinder involved balancing a requirement to technically progress the systems development with business and programme implementation support requirements.

The aims were to :

• ensure that there was a convergence of the activities of the Pathfinder team

- provide a full assessment of the vendor system and support capability
- identify how to optimise benefits and minimise the risks for the business.

In addition to demonstrating the compliance of the selected package to meet the VSEL requirements, other key objectives during Pathfinder included:

- the development of a robust **implementation plan** that addressed the roll out of the hardware, software, communications, people and training aspects
- a set of **metrics** that provided confidence in the **benefit estimates**
- a group of trained and aware users that formed the **core champions** for the deployment of CAD and EDM in the project areas
- a **management team** that was aware and supportive of the practical aspects of change necessary to implement CAD and EDM
- a management team that was ready to support the **people** and **processes** issues involved in the roll out through pilots and by application to projects. A clear linkage was made between TO BE processes and the use of CAD /EDM
- an **organisation** that can logically be developed from the Pathfinder to take the implementation of CAD and EDM into the next phase with trained and motivated individuals ready to fill key roles

5 CONCLUSIONS AND LESSONS

VSEL's interest in implementing a Concurrent Engineering approach was driven by very real pressures and clearly defined needs. Maintaining a technical capability that would differentiate VSEL in the competitive market place was a survival issue. Linked to this was the need to continue to win new business and increase the profitability of the company.

The response was a Business Improvement Programme with the Effective Management of Design as a major initiative. The position of the design activities at the front end of the development process generally means that some 80% of the commitments to the eventual total cost of the product will be made during the expenditure of the initial 20% of the costs. Many of the opportunities to change and influence the process and generate beneficial improvements thus start at the design stage and manifest themselves in downstream processes such as Procurement, Production and Support.

To release these benefits, it was recognised that changes in People, Processes and Systems would be involved. It was the combination and integration of changes in these three areas that have provided the Concurrent Engineering model for VSEL.

The systems and technology elements offered extensive opportunities to realise new and streamlined processes and to improve the efficiency of individual tasks.

This new generation toolkit of supporting tools includes: *Electronic Product Definition* for early evaluation of designs, realistic, lifelike *Visualisations* of solutions that dramatically improve communications and *Engineering Data Management* system capabilities that can manage all of the data required to define a product or execute a contract.

A technology demonstrator giving a practical illustration of how these opportunities can be exploited has 'set the agenda' for the future way of working.

The main language for discussing the changes has been the process model. The starting point was the current process or AS IS and the analysis of all the associated 'sore thumbs', gaps, issues, bottlenecks and opportunities. The TO BE model includes activities to create the CE environment and resolve the problem areas.

Change Mechanisms (or the things you have to do to bring about a new way of working) have been defined together with the process whereby they are applied to improvement themes to create benefits in defined areas through the use of pilot projects.

A series of pilot projects were launched, starting with the Pathfinder. This was the route to convert the theoretical TO BE model into a series of practical experiences for a widening group of trained team members. Pathfinder was aimed specifically at reducing the risks associated with the CAD and EDM systems and provided the model for preparing, launching and undertaking pilot projects.

Positive evidence of the benefits is building as implementation continues, with the pilots producing improvements in:

Design Productivity :
- Re-use of standard part data
- Improved efficiency and effectiveness of Process
- Timely, accurate and controlled information release
- Reduced duplication of effort, particularly with data input
- Less rework in design
- Single source of controlled product data

Procurement Productivity:
- Improved and accurate material definition
- Reduced inventory through better planning
- Less material scrap due to late design change
- Electronic Data Interchange with suppliers

Production Productivity :
- Reduction in rework & scrap
- Right information at the right time
- Material definition linked to work packages
- Cost of Change identified and managed
- Improved and Integrated Planning
- Improved Build Strategy
- Design for Production

Support Productivity :
- Single Product Data Model
- Logistic Support Analysis Record
- Configuration Control and Management
- Data availability
- CALS compliant data exchange

A risk assessment of the project was undertaken and identified the dangers of attempting to undertake too many changes at one time and the need to be realistic about what can be achieved in time to influence ongoing projects. Other risks identified related to the shortage of suitable resources to implement the system and process changes, the difficulties associated with resistance to change and the level of effort required to overcome inertia or disbelief in the proposed changes.

The process changes and systems issues were difficult but could be quantified and addressed in a systematic way. Success in addressing the people issues was key in determining the level of benefits. Training, communication and 'experience by doing' has been the main approach used. This progressively influences the behaviour of the individual as they are actively involved in the changes. Success of an initiative or pilot then creates positive attitudes towards change and reinforcement of this progressively develops the new way of working to become the normal way of operating.

From the experiences of implementing Concurrent Engineering at VSEL, the following are the things you must do to ensure success:
- ❏ obtain Senior Management commitment and support

- ❑ re-design processes - 'as is' and actions to move to a 'to be'
- ❑ achieve the maximum level of controlled parallel activities using a 'tollgate' approach and 'pull planning'
- ❑ assign single point responsibilities
- ❑ provide the authority to fulfil the responsibilities (empower people)
- ❑ modify behaviours, change attitudes and develop a culture of mutual trust and respect
- ❑ ensure that all inputs are simultaneously present when decisions are taken
- ❑ design the product in parallel with determining the means of production and the support needs (Design Build teams)
- ❑ use effective techniques to assess the impact of actions or decisions on the 'whole product lifecycle and design'
- ❑ be able to handle the incremental development of information (partial data)
- ❑ provide visibility of information and management of change through supportive systems
- ❑ communicate 'til it hurts !

No one said it would be easy !

ACKNOWLEDGEMENTS:

The authors, Stephen Foster and Barry Brooks would like to thank Vickers Shipbuilding and Engineering Limited, Barrow-in-Furness and PA Consulting Group, Cambridge, for permission to publish this case study.

Index